SpringerBriefs in Computer Science

Series Editors
Stan Zdonik
Peng Ning
Shashi Shekhar
Jonathan Katz
Xindong Wu
Lakhmi C. Jain
David Padua
Xuemin Shen
Borko Furht
VS Subrahmanian
Martial Hebert
Katsushi Ikeuchi
Bruno Siciliano

T0214487

For further volumes:
http://www.springer.com/series/10028

Hao Zhang • Yonggang Wen • Haiyong Xie
Nenghai Yu

Distributed Hash Table

Theory, Platforms and Applications

 Springer

Hao Zhang
Department of Electronic Engineering
  and Information Science
University of Science and Technology
  of China
Hefei, Anhui, China

Haiyong Xie
Department of Computer Science & Suzhou
  Institute for Advanced Study
University of Science and Technology
  of China
Suzhou, Jiangsu, China

Yonggang Wen
School of Computer Engineering
Nanyang Technological University
Singapore

Nenghai Yu
Department of Electronic Engineering
  and Information Science
University of Science and Technology
  of China
Hefei, Anhui, China

ISSN 2191-5768   ISSN 2191-5776 (electronic)
ISBN 978-1-4614-9007-4   ISBN 978-1-4614-9008-1 (eBook)
DOI 10.1007/978-1-4614-9008-1
Springer New York Heidelberg Dordrecht London

Library of Congress Control Number: 2013949599

Printed on acid-free paper

Springer is part of Springer Science+Business Media (www.springer.com)

Preface

Distributed hash table (DHT) has been playing an important role in distributed systems and applications, especially in large-scale distributed environments. DHT was introduced to address a daunting challenge in large-scale system architecture. Specifically, in a normal client/server model (C/S model), centralized servers would potentially become the bottleneck of the whole system. As a comparison, the distributed model, exemplified by the peer-to-peer (P2P) model, leverages the resources spread across a list of nodes in the system. At the same time, it is desirable to utilize all the peers' capability efficiently and provide better robustness. DHT technology was developed to meet these requirements. Indeed, in DHT, distributed resources are managed so well that peers only need to know part of the system. The elegance of DHT is its implicity in operations, providing only two basic operations, including: (i) GET data from DHT and (ii) PUT data into DHT. Finally, given its simplicity, DHT is yet suitable for a great variety of applications and provides robustness and high efficiency, especially in large-scale systems.

For decades extensive work has been conducted for DHT. In academia, researchers have proposed several variants of DHT and associated improvements, which manage the resources in different structures, providing abundant choices to build distributed systems. Meanwhile, many practical platforms of DHT have been implemented, which can be regarded as a bridge translating DHT from theory to practice and solving many practical problems such as load balance, multiple replicas, consistency, latency, and so on. Finally, a lot of applications based on DHT have been proposed, for example, multicast, anycast, distributed file systems, search, storage, content delivery network, file sharing, and communication. Previous surveys on DHT have been mainly focused on the theoretic aspect, with less attention paid to platforms and applications.

In this book, we aim to report the development of DHT in both academic pursuit and industrial development. It covers the main theory, platforms, and applications of DHT. From this book, readers could learn the basic principle of several popular DHT structures, many platforms used in both academic and commercial fields, and a wide range of DHT-based applications. We have also presented our view of potential limitations of DHT.

This book consists of five chapters. In Chap. 1 background information about DHT is introduced. Seven variants of DHT are studied and compared in Chap. 2. In Chap. 3, we classify 15 existing DHT platforms into two categories: (i) academic and open-source platforms and (ii) commercial platforms. In Chap. 4 we present eight DHT-based applications with detailed analysis of their pros and cons. In Chap. 5, we outline the benefits and limitations of DHT.

Heifei, China   Hao Zhang
Singapore, Singapore   Yonggang Wen
Suzhou, China   Haiyong Xie
Heifei, China   Nenghai Yu

Contents

Chapter 1
Introduction

Nowadays distributed hash table (DHT) [1, 2] plays an important role in distributed systems and applications, especially in large-scale distributed environments. In the normal Client/Server model (C/S model) since the central server is in charge of most of the resources, it becomes the most important part as well as the bottleneck and weak point of the system. On the contrary, the distributed model (a typical one is the peer-to-peer (P2P) model [3,4]) distributes the resources on the nodes in the system. The distributed model provides better robustness and more efficiently utilizes all peers' capability, while the resources of the clients are idle in C/S mode. In distributed environments a key problem is how to manage the resources efficiently, which is a particular important issue in large-scale systems. DHT addresses this problem and promotes the development of P2P greatly.

DHT is a simple and elegant design for distributed systems. It provides the functions like a hash table to deal with the distributed data. DHT does not require a central server and treats all DHT nodes in the distributed system equally. Meanwhile, DHT inherits the great properties of hash table (e.g., locate and search an element with high efficiency). DHT provides a global, abstract key space (often referred to as the DHT space), where all resources (e.g., data and DHT nodes) have unique identifiers (IDs). Like in the hash table, any data in DHT could be treated as a tuple (K, V), where K denotes the key that is mapped from the data by a hash function and V denotes the original data. Each node also has a key called ID of the node in the DHT space. Thus all data and nodes in a distributed system can be consistently mapped into the DHT space. The DHT space is split into slots; each node in a DHT system maintains the data that are mapped into this node's slot. As a result of its simple and elegant design, DHT has two primitive operations: put () is a function that puts data V into the DHT space with a key K. get () is a function that gets the original data using a given key K. Although extremely simple, these two primitives are suitable for a great variety of applications and provide good robustness and high efficiency, especially in large-scale systems.

DHT organizes the distributed resources so well that nodes only need to know a part of the system from which they can get resources efficiently, and resources can

H. Zhang et al., *Distributed Hash Table; Theory, Platforms and Applications*,
SpringerBriefs in Computer Science, DOI 10.1007/978-1-4614-9008-1__1,
© The Author(s) 2013

be located in $O(1)$ time regardless of how many resources are in the space (here we do not consider the cost of underlying network routing). Furthermore, DHT is more capable of dealing with system distributivity and dynamics than the regular hash tables, since DHT can better adapt to the varying number and range of slots. It possesses the following three properties:

1. **High efficiency**. DHT inherits the excellent properties of hash table, e.g., efficiently locating where data are stored in the DHT space without knowing the global information. Efficiency is a very important concern in distributed systems. Moreover, without knowing the global information means that every node in a DHT system needs to know only a part of the system and works with others cooperatively. This property is especially important and desirable for large-scale systems.
2. **Decentralization.** DHT is not deployed on a single node but on many ones, each of which maintains a portion of the hash table. This property means that there is no central node, which could avoid the hot spot problem and achieves a good load balance.
3. **Scalability.** This property is an outcome of the decentralization property. DHT can be applied to distributed systems with varying sizes, ranging from several, thousands of up to millions of nodes. However, the events of node joining and leaving the system as well as node failures are not uncommon in such systems, which means that DHT should handle these problems efficiently.

Consistent hashing [5] can be viewed as an early version of DHT and was proposed in 1997 for distributed caching systems. In distributed caching systems, regular solutions to mapping content objects to caching servers are typically based on the operation $h(object) \mod N$, where $h()$ is a hash function specifically chosen for individual system, and N is the total number of caching servers. However, the dynamics of such systems (e.g., removing a server due to failure or adding a new server) require content objects to be re-mapped to $N - 1$ (or $N + 1$) regular caching servers by $h(object) \mod (N - 1)$ [or $h(object) \mod (N + 1)$]. Such re-mapping operations require every regular server to refresh the data that it maintains. It costs too much when N is large. Generally speaking, node removal and addition are not uncommon in large-scale distributed systems. This requires a dynamic hash method to migrate data with affordable costs, e.g., a hash method with good monotonicity [5]. Consistent hashing is a kind of hash algorithm that maps all the keys and cache servers on a circle by common hash function, where each server maintains the data on an arc (called slot) of this circle. In this way, consistent hashing provides hash functionality in a way that it supports addition or removal of slots by splitting arc or merging the neighbor arcs. Only a few servers have to update the objects stored in their caches. Nowadays consistent hashing is widely used in distributed caching systems. Not only in the caching scope, this architecture is also borrowed by some of the P2P systems such as Chord [6], which encourages the adoption of DHT greatly. For example, according to Google Scholar, the seminal paper on Chord has been cited by more than 10,000 times by the end of 2012. This suggests that DHT has been a popular research topic especially in P2P networks.

In the past decade, extensive work has been done for DHT. In academia, researchers have proposed numerous variants of DHT and improvements, which manage the resources in many kinds of structures, providing abundant choices for the construction of distributed system. Meanwhile, many platforms of DHT are constructed, which can be regarded as a bridge transforming DHT from theory to practice. They solve many practical problems such as load balance [7, 8], multiple replicas [9, 10], consistency [11–13], latency [9, 13], security [14, 15] and so on. Furthermore, lots of applications based on DHT are proposed such as multicast, anycast, distributed file systems, search, storage, content delivery network, file sharing and communication. There are also some surveys on DHT, most of which only focus on theory of DHT. In [16] several DHT theories are described. In [17, 18] DHT is introduced as a part of P2P networks. In [19] the authors introduce the theory of DHT from another aspect, where all the variants of DHT are classified by their topologies. However, in these surveys the authors ignored many of DHT platforms and applications that are very important, especially in the industrial area.

This book includes five chapters. In Chap. 1 some background information about DHT is introduced. Seven variants of DHT are studied and compared in many aspects in Chap. 2, which is the basis of the platforms. In Chap. 3 two kinds of platforms (academic and open-source platform and commercial platform) containing 15 different platforms are analyzed, based on which applications can be constructed. In Chap. 4 we present eight application scenarios, and the advantages of DHT in these applications. Chapter 5 summarizes the book by discussing the pros and cons of DHT.

Chapter 2
DHT Theory

In this chapter, we discuss a set of DHT variants, which widely influence the design and development of distributed systems in recent years. In each section, we first describe the structure of each variant, present the key elements such as the routing model and the data model, and then discuss the solutions to a common problem in distributed environments [20], namely, nodes dynamically join and leave distributed systems. In the last section, we compare the DHT variants from numerous aspects, such as overlay network topology, distance metric, routing and data model and so on.

2.1 Chord

Chord [6] is a distributed lookup protocol. It solves a general but fundamental problem in P2P networks, namely, how to efficiently locate the node which stores a particular data item. In addition, Chord is designed to address challenges in P2P systems and applications, for instance, load balancing, decentralization, scalability, availability and flexible naming.

In Chord, the DHT space is a circle, which is a one-dimensional space referred to as the *Chord ring*. Although the structure of Chord is a ring-like consistent hashing, their aims are different. Consistent hashing focuses on the caching problem, while Chord is an architecture for organizing the nodes and contents in P2P networks. In Chord, both nodes and data are mapped into this space by a pre-determined hash function (e.g., SHA-1 [21]). The keys used to map nodes and data are referred to as the identifiers (IDs) of the corresponding nodes and data. IDs for nodes can be generated by applying a hash function to unique information of individual nodes (e.g., nodes' IP addresses), and IDs for data can be computed by applying a hash function to the data themselves.

IDs are ordered on the Chord ring by calculating ID mod 2^m, where m is the number of bits in the key. Therefore, IDs correspond to points on the Chord ring. All IDs are arranged clockwise in an ascending order on the Chord ring. For a node N_i with its ID being i, we define its previous node on the clockwise ring as

H. Zhang et al., *Distributed Hash Table; Theory, Platforms and Applications*,
SpringerBriefs in Computer Science, DOI 10.1007/978-1-4614-9008-1_2,
© The Author(s) 2013

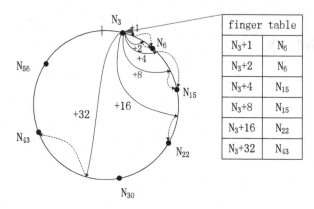

Fig. 2.1 An example of finger table

$\texttt{predecessor}(N_i)$, and define its next node as $\texttt{successor}(N_i)$. In particular, the node with the maximal ID (i.e., $N_{ID_{max}}$) chooses the node with the minimal ID (i.e., $N_{ID_{min}}$) as its successor. Additionally, $N_{ID_{max}}$ is the predecessor of $N_{ID_{min}}$.

However, if each node only knows its predecessor and successor in the one-dimensional Chord ring (which is a directed graph), such a DHT system would be inefficient and vulnerable for numerous reasons. First, the time complexity of DHT lookup is $O(n)$, where n is the number of peers in such a system; hence, the complexity could be too high when n is large. Second, each node can only send messages to its successor on the ring; hence, the node connectivity is 1 in such a system (both in-degree and out-degree). Third, since each node has only one choice for routing, if one node fails, the connectivity of the graph will be destroyed.

Chord introduces a *finger table* structure to solve the above problems. Each node has a finger table, and each finger table maintains up to m nodes (recall that the size of the key space is 2^m) for the purpose of efficient routing and increasing the connectivity of the graph. For an arbitrary node N, the ith entry in its finger table contains the first node clockwise from $N + 2^{i-1}$:

$$\texttt{N.finger}[i] \quad = \quad \texttt{successor}(N + 2^{i-1}). \tag{2.1}$$

Each node maintains the data whose IDs are in the range between this node's ID and its predecessor's ID. The finger table structure improves the connectivity of the Chord ring and thus improves the routing on the ring, which in turn significantly reduces the complexity. More specifically, the time complexity of the DHT lookup operation is reduced from $O(n)$ to $O(\log n)$ due to increasing the connectivity from 1 to $O(m)$.

Figure 2.1 shows an example of the Chord ring where $m = 6$ and the finger table of node N_3. The first entry (i.e., the zeroth entry) in this finger table contains N_6, since according to Eq. 2.1, the ID of the first neighboring node defined by the finger table should be $(3 + 2^0) \mod 2^6 = 4$. However, there does not exist a node whose ID is 4. Therefore, the immediate next node on the ring, i.e., N_6, is chosen for the reason that successor(4) = N_6. Similarly, the last entry (i.e., the fifth entry) in the finger table contains N_{43}, because $(3 + 2^5) \mod 2^6 = 35$ and the first node that succeeds the node N_{35} is successor(35) = 43.

The two primitives for data storage and retrieval are as follows: (1) put (k, v) stores a given data v whose ID is k on the node that has an ID closest to k, and (2) v = get(k) retrieves the corresponding data v stored at a node using the ID k. The common key to both data storage and retrieval is how to locate the node that is responsible for storing the data with a given ID. Chord locates the node as follows. When a node N with ID j needs to locate where the data with ID k is stored (or should be stored), it would send a query to the node N' satisfying the Eq. 2.2:

$$N' = \begin{cases} \text{N.finger}[0] & \text{d}(k, j) \leq \text{d}(\text{N.finger}[0], j) \\ \text{N.finger}[i] & \text{d}(k, j) \leq \text{d}(\text{N.finger}[i + 1], j) \text{ and } \text{d}(k, j) > \text{d}(\text{N.finger}[i], j) \\ \text{N.finger}[m - 1] & \text{otherwise} \end{cases}$$

(2.2)

In Eq. 2.2, N.finger[0] is the ID of node N, and the distance between ID x and ID y is $\text{d}(x, y) = (x - y) \mod 2^m$, which is the Euclidean distance in the one-dimensional ring space. Note that during data storage and retrieval, each node always tries to send queries to the node that is the closest to the querying node.

A node may dynamically join or leave a Chord system. The primitive join() inserts a new joining node into the Chord ring and updates relevant nodes' successors accordingly. When a node joins a Chord system, it is assumed to know at least one node on the Chord ring, which helps the new node to locate its successor. Moreover, a stabilization protocol runs periodically to update the successor lists and finger tables. The primitive leave() removes a voluntarily leaving node from the Chord ring and updates the lists of successors and finger tables accordingly. When a node leaves or fails, some other node may lose its successor (if the leaving/failing node is the successor). To mitigate this situation, each node maintains a list of the first r successors. When one of its successors leaves or fails, a node simply chooses the next node on this list as the successor. By tuning the parameter of r, Chord could balance the robustness and the cost of maintaining the successor list.

Researchers have studied how to improve Chord extensively and there has been a large body of literature on Chord; To name a few of such studies, Flocchini et al. [22] proposed a method that combines multiple Chord rings to achieve data redundancy and reduce the average routing path length; Joung et al. [23] proposed a two-layer structure called $Chord^2$, where super peers are introduced to construct a conduct ring which could reduce the maintenance cost; Kaashoek et al. introduced a Chord-like structure Koorde [24] where the bruijn graphs [25] substitute the finger table; Cordasco et al. [26] proposed a family of Chord-based P2P schemes, F-Chord(α), using the Fibonacci numbers to improve the degree, diameter and average path length of Chord. Furthermore, H-F-Chord(α) using the NoN (Neighbors of Neighbors) technique [27–29] is more efficient in terms of its average path length $O(\log n / \log \log n)$; Ganesan et al. [30] optimizes Chord routing algorithms by exploiting the bidirectional edges, both clockwise and counterclockwise, which reduces the average routing path length.

2.2   Content-Addressable Network (CAN)

CAN [31] is a distributed, Internet-scale, DHT-based infrastructure that provides hash table-like functionalities. Different from the one-dimensional space in Chord, the DHT space in CAN is a d-dimensional Cartesian space. The d-dimensional space is further dynamically partitioned among all nodes and each node only maintains its own individual and distinct zone in the space.

In CAN, every node maintains $2d$ neighbors, thus the node connectivity is $2d$. Here the notion of "neighbors" means two zones that overlap along $d-1$ dimensions and have neighbors on one dimension. As a result, CAN dose not need to introduce complex structures such as long links (e.g., the finger table in Chord) connecting nodes, which are further away from each other in the d-dimensional space, in order to improve the connectivity and reduce the complexity of routing. Note that d is a constant independent from the number of nodes in the system, which means that the number of neighbors each node maintains is a constant, no matter how many nodes the CAN system may have.

Figure 2.2 illustrates a two-dimensional CAN with 20 nodes. Each dimension covers $[0, 1)$ and every node maintains a zone in the grid. For example, node 1 maintains the zone $(0 - 0.25, 0.75 - 1.0)$, and node 17 maintains the zone $(0.375 - 0.5, 0.5 - 0.75)$. Every node maintains the IDs (e.g., IP addresses) of its neighbors that maintain the zones in the neighborhood.

The routing in CAN works as follows. When receiving a message with a specific destination, a node routes the message towards the destination using a simple greedy algorithm, i.e., the node goes through the list of its neighbors to select the one that is closest to the destination, and then forwards the message to the selected neighbor.

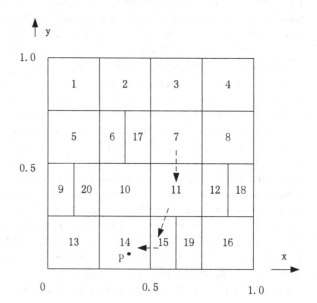

Fig. 2.2 Example of a two-dimensional CAN with 20 nodes

This greedy forwarding process continues until the message arrives to the designated destination. For instance, Fig. 2.2 shows a routing path from the node 7 to the point P in the zone maintained by the node 14. The dashed lines illustrate the steps in which nodes greedily forward a message from the source (i.e., node 7) to the destination P.

The data in CAN is stored and retrieved based on the notion of key-value pairs, similar to the notion adopted in Chord. More specifically, in order to store a key-value pair (k, v) where k is the key and v is the corresponding value (i.e., data), k is first mapped to a point P in the d-dimensional space using a hash function specifically chosen by CAN; then the corresponding (k, v) pair is delivered to and stored on the node that maintains the zone where P is located. Similarly, in order to retrieve the value v for a given key k, a node should first obtain the point P by mapping the key k, and then retrieve the corresponding value v from the node that maintains the zone where P is located.

The protocol that accommodates the dynamic node arrival and departure is more complex than that in Chord, due to the fact that the d-dimensional space adopted by CAN is more complex than the one-dimensional space by Chord. When a node i joins the CAN system, it should be introduced into the system by another node j that is already in the system. More specifically, the new joining node must send a JOIN request to find an existing node whose zone z_j can be split. This zone should be split by a certain number of dimensions, in such a way that the zone can be reclaimed when nodes leave in the future. When a node i leaves the system, the zone z_i that it maintains should be taken over by other remaining nodes. If the zone z_i can be merged with a zone z_j which is maintained by a node j (one of the neighbors of the leaving node i), then the node j should reclaim the zone z_i by merging it with its own zone z_j and maintain the new larger zone. Otherwise, the zone z_i should be taken over by the neighbor whose zone is the smallest.

For instance, suppose that the first splitable dimension is the x axis. When a new node (ID is 21) is joining the system, it finds that the zone maintained by the node 2 can be split in half. After the split, the zone maintained by the node 2 changes from $(0.25 - 0.5, 0.75 - 1.0)$ to $(0.25 - 0.375, 0.75 - 1.0)$. The other half, i.e., $(0.375 - 0.5, 0.75 - 1.0)$, will be assigned to the joining node 21. When the node 17 leaves, its zone should be reclaimed and merged with the zone maintained by the node 6. Note that after the zones split (and merge), the neighbors of the affected zones should update their neighbor lists so that the joining node (and departing node) can participate in (and be removed from) the routing system. For example, before the node 17 joins in the CAN system, the neighbor list of the node 6 is $\{2, 5, 7, 10\}$ and the list of the node 7 is $\{3, 6, 8, 11\}$. After the node 17 successfully joins, the zone maintained by the node 6 is split into two smaller zones, one is maintained by the node 6, the other is by the node 17. Then node 6, 17 and the neighbors 2, 5, 7, 10 should update their neighbor lists. After updating the neighbor lists, the neighbors of the node 6 are $\{2, 5, 10, 17\}$, and the neighbors of the node 7 are $\{3, 8, 11, 17\}$.

To maintain the healthiness of the CAN system, each node periodically sends update massages to its neighbors. When a node has not received the update massages from one of its neighbors for a long time (longer than a certain timeout threshold), it considers that the neighbor fails and starts a takeover timer for it. When the timer expires, the node sends a TAKEOVER message to all neighbors of the failed nodes, announcing that it takes over the zone that is formerly maintained by the failed node. However, if the node receives a TAKEOVER message before the timer expires, the node then cancels this timer. When such a timer is initialized, it should be set up in a way that it is proportional to the failed node's zone. More specifically, if one node fails, the neighbor who maintains the smallest zone should send the TAKEOVER massage at the earliest moment and therefore takes over the failing node's zone.

Besides, to improve the robustness and performance of the CAN system, three technologies, i.e. increasing dimensions, multiple realities and RTT-weighted routing, are introduced [31]. As a result, CAN behaves well in large-scale distributed systems and has been applied to many applications in such systems.

2.3   Global Information Sharing Protocol (GISP)

GISP [32] is a protocol for fully decentralized P2P networks. GISP does not make any assumption on the network structure, thus it is applicable to both structured and unstructured P2P networks.

There are a set of key principles that guide the design of GISP. First, each node should maintain as much peer information as possible, so the network has great connectivity. Second, each node should discard information of unreachable peers and keep more information about nodes that are numerically closer. Last but not least, each node may possess different levels of capability; however, the more capability one node possesses, the greater responsibility the node should take. GISP introduces the notion of "peer strength" to quantify the capability of a node. The powerful nodes (with higher peer strength) should keep more data and have more connectivity than those with less peer strength.

The routing model in GISP works as follows. GISP leverages hash functions such as MD5 and SHA-1 to map any binary data including a keyword into a number with fixed bit length. Similar to Chord and CAN, each node in GISP has a unique ID in the hash space. GISP defines the distance between two nodes by $distance(i, j)/(2^{s_i-1}2^{s_j-1})$, where i and j are the IDs of two nodes, s_i and s_j are the values of the two nodes' "peer strength." GISP adopts a greedy routing strategy, namely, when forwarding a message to its destination, a node selects the next-hop node who has the shortest distance to the destination.

The data storage and retrieval are similar to Chord. When inserting the data v into a GISP system, the key k is computed (i.e., the hash value of the data v) and the node whose ID is numerically closest to k is selected to maintain the data. In other words, GISP selects the node who has the shortest distance to the hash value of the data. For example, suppose that a GISP system has four nodes, whose ID are 100, 110, 115, 122 respectively. If a piece of data with key 107 is pushed in the system, then the node with ID is 110 would be selected to maintain the data.

Data retrieval is similar to the data storage process. More specifically, given a key k, the node that is responsible for maintaining the data is located and queried, then the data v will be routed back to the requesting node. Data storage and retrieval are conceptually straightforward; however, it is difficult to completely delete any stored data. In GISP, when a node stores a piece of data, it also set up a timer for the data; each node periodically check the data it maintains and delete the expired data.

Since node failures in P2P networks are not uncommon, in GISP the data is duplicated and the replicas are distributed to multiple nodes whose IDs are closer to the hash value of the corresponding data. The number of data replicas is determined either statically or dynamically. The larger the number is, the more robust the system is to multiple node failures, and the more storage capacity is required. As a result, routing messages are not sent to only one node but to a group of peers (in the ascending order of the distance from the nodes to the hash value of the data). In order to avoid routing loops, each message is associated with a list of peers to which this message has already been sent. For example, suppose that the node 4 is sending a message M to the nodes 5, 6 and 7. When node 4 sends M to node 7, it also tells node 7 that it has already sent this message M to node 5 and 6. Thus, node 7 will not route M to 5 and 6.

When a node i is joining a GISP system, it is assumed that node i knows at least an active node j in the system. From the node j, node i acquires knowledge about other nodes in the network. When a node i leaves the system, it notifies other peers its departure. In the case that some nodes leave the system without notifying others (e.g., due to the network connectivity problems), GISP can still work well unless too many peers fail at the same time and as a result too much data stored in the network is lost.

GISP introduces a latency-based mechanism to relate the overlay network topology with the real underlying physical network topology. Such a mechanism can reduce the cost of network routing in GISP. More specifically, when a new node is joining the network, GISP first determines this node's ID based on the latency values of the existing nodes that it knows. By doing so, nodes that are closer in the underlay network topology are likely to form clusters (i.e., the distances/latencies between nodes in such clusters are lower) in the overlay network topology.

2.4 Kademlia

Kademlia [33] is a P2P storage and lookup system. Both nodes and data in Kademlia are assigned with 160-bit integer IDs. More specifically, each node chooses a random 160-bit integer as its ID, and data is stored in the form of key-value pairs, where the key is a 160-bit value generated by hash functions such as SHA-1 and being the ID of the value, and the value is the data stored in Kademlia.

Unlike Chord and CAN, Kademlia defines the distance between two nodes i and j by the bitwise exclusive OR operation (XOR), i.e., $d(i, j) = i \oplus j$. This distance metric is unidirectional like the metric used in Chord, which means for any given key i and a distance $l > 0$, there are only one key j that satisfies $d(i, j) = l$.

Every piece of data in the form of key-value pair is stored on k nodes whose IDs are the closest to the key. Here k is a key parameter in Kademlia to determine data redundancy and system stability.

Each node i in Kademlia maintains multiple k-buckets. Each k-bucket is a linked list with a maximum length of k. Each k-bucket keeps a list of nodes, which are sorted in the ascending order of recent activities, i.e., the node that is the least recently seen is stored at the head, and the node that is the most recently seen is stored at the tail. The node whose distance from the node i is in the range of $[2^m, 2^{m+1}]$ is stored in the mth k-bucket (note that $0 \le m < 160$). The nodes in the k-buckets are regarded as the neighbors of the node i.

Unlike in Chord and CAN, a node updates its neighbors dynamically upon receiving any messages from them. More specifically, when a node i receives a message from another node j, which is located in the mth k-bucket, this k-bucket of node i will be updated in the following way. If j already exists in the k-bucket, i moves j to the tail of the list, as node j is the node that is the most recently seen. If j is not in the k-bucket and the bucket has fewer than k nodes, node i just inserts j at the tail of the list. If the bucket is full, i pings the node at the head of this k-bucket. If this head node responds, node i moves it to the tail and ignores node j. Otherwise, i removes the head node and inserts j at the tail.

Kademlia has four RPC-like primitives, i.e., PING, STORE, FIND_NODE and FIND_VALUE. The PING primitive probes a node to check whether it is online or not. The STORE primitive is used to store a key-value pair. The FIND_NODE primitive finds a set of nodes that are closest to a given node; in other words, it returns k nodes from one or multiple k-buckets, whose IDs are closest to the given node's 160-bit ID. The FIND_VALUE primitive behaves like FIND_NODE, except that it returns the stored value. These primitives work in a recursive way, and in order to improve the efficiency of Kademlia, a lookup procedure is invoked by the FIND_NODE and FIND_VALUE primitives. More specifically, at the beginning, the lookup initiator picks α nodes from its closest k-bucket and sends multiple parallel FIND_NODE requests to these α nodes. If the proper node is not found, the initiator re-sends the FIND_NODE to the nodes it just learned in the last recursive execution. A key-value pair may be stored on multiple nodes. With the recursive lookup procedure, the key-value pair spreads across the network every hour. This method ensures that for any data, multiple replicas exist for robustness. Every key-value pair is deleted 24 h after it is initially pushed into the network.

When a node i is joining the network, it is assumed that it knows a node j which is active and already in Kademlia. The joining process consists of multiple steps. First, node i inserts j into its k-buckets. Second, node i starts a node lookup procedure for its own ID, from which i learns some of the new nodes. Finally, node i updates the k-buckets. During this process, node i strengthens its k-buckets and inserts itself into other nodes' k-buckets. When a node fails or leaves, it does not notify any other node. There is no need for a special procedure to cope with node departures, as the mechanism of k-buckets ensures that the leaving nodes will be removed from the k-buckets.

For the reasons of the simple distance metric and the k-bucket mechanism, Kademlia becomes the most widely used DHT system—it has been adopted by many popular P2P applications such as Overnet [34], eDonkey™/eMule™ [35] and BitTorrent™ [36, 37]. Researchers have made many efforts on analyzing and improving the lookup performance of the Kademlia protocol [38–42], in order to enhance the practicality of the Kademlia-based system.

2.5 Pastry

Pastry [43] is a self-organizing overlay network with a targeted basic capability of routing messages efficiently. Every node in Pastry has a 128-bit ID. A node ID is divided into multiple levels, each of which represents a domain. Each domain is represented by b (an integer by which 128 is divisible) contiguous bits in the node ID, i.e., the domain at level l is specified by the bits at positions $b \times l$ to $b \times (l+1) - 1$. Each level contains 2^b domains numbered from 0 to $2^b - 1$. Figure 2.3 illustrates an example of dividing the Pastry node ID with $b = 4$. The first four bits specify the domain at level 0, and the following four bits specify the domain at level 1. In this case the domain at level l is domain 9 (i.e., the binary bit string is 1001).

Routing Model

Each node has a routing table, a neighborhood set and a namespace set. The routing table of a node contains $2^b - 1$ nodes for each level l; these nodes have the same prefix up to level $l - 1$ as the local node. Hence, the routing table contains $L \times 2^b - 1$ nodes, where L is the number of the levels. The neighborhood set contains M nodes, which are closest to the local node (measured by their physical distances). However, note that the neighborhood set is not used in routing messages. The namespace set contains L nodes which are closest to and centered around the local node. The namespace set is used during the message routing and object insertion.

When a node routes an incoming message, the node first checks if the destination's ID falls in its namespace set. If so, the message will be sent directly to the destination node. Otherwise, the node uses the routing table to choose the domain at a level l, where the nodes at this level l share the longest prefix with the destination node's ID. Then the node selects a node in this domain as the next hop. The selected node has to be alive and be closer to the destination than other nodes in the same domain.

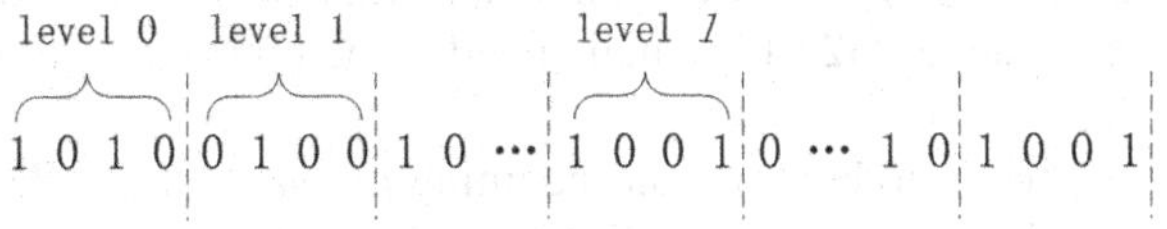

Fig. 2.3 Node ID division in Pastry with b = 4

Data Model

Every data object v in Pastry has an object ID k that is at least 128 bits long, which can be generated by a hash function. When storing an object v into the Pastry network, Pastry routes a message containing the data object v to the node whose ID is numerically closest to k (i.e., the ID of v). In order to improve the data availability, each data object is not only stored on one node but also on a set of extra nodes whose IDs are numerically close to the object ID.

When a node i joins a Pastry system, it is assumed that the node knows at least an active node j in the network. Node j routes a "join" message on behalf of node i and the message is destined for node i. Eventually, the message will be routed to a node i' whose ID is numerically closest to i. Then node i copies node j's neighborhood set as its initial neighborhood set, and takes the namespace set of node i' as its initial namespace set. Node i also initializes its routing table using the relevant information of the nodes on the path from j to i', including j and i'. When a node i is leaving a Pastry system, there is no specific protocols to handle the node departure. Rather, nodes solve this problem by refreshing their routing table, neighborhood set and namespace set.

2.6   Tapestry

Tapestry [44] is a P2P overlay routing infrastructure which offers scalable, location-independent and efficient routing of messages using only local resources.

In Tapestry, each node has a 160-bit ID, and each application-specific endpoints (e.g., objects) is assigned with a 160-bit globally unique identifier (GUID), both of which can be generated by using a hash function such as SHA-1. The "distance" between two nodes is assessed digit by digit; for example, a node whose ID is "2341" which is closer to the node "2347" than the node "2338." Below, we will elaborate the routing and data model of Tapestry, respectively.

Routing Model

Each node maintains a routing table which consists of a set of neighboring nodes. Tapestry guarantees that any node can be reached from the root in at most $\log_\beta N$ hops, where N is the size of the name space and β is the base of IDs. In order to do so, neighboring nodes in the routing table are organized into $\log_\beta N$ levels. In the jth level, at most $c \times \beta$ pointers to the relative nodes that begins with prefix($N, j - 1$) are maintained, where c neighbors differ only on the jth digit. For instance, "$325A$" is maintained in the fourth level of the routing table of "3259" with $\beta = 16$.

When a node routes an incoming message, the node selects the next-hop node from the neighbors by matching the level prefix, which is similar to the longest

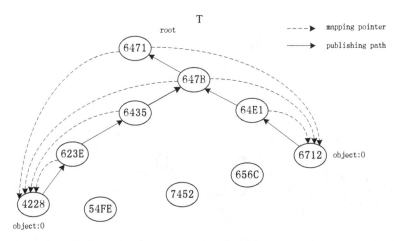

Fig. 2.4 An example of storing (publishing) a data object in Tapestry

prefix routing method used in the CIDR IP address allocation architecture [45]. As a result, the IDs of the nodes on a route vary gradually (e.g., 3*** $\Rightarrow$ 34** $\Rightarrow$ 34A* $\Rightarrow$ 34AE, where "*" is the wildcard). Tapestry also provides the surrogate routing to route messages to some active nodes with similar IDs. This scheme can mitigate the problem of single-point failures.

Data Model

Each object in Tapestry has a root, as Tapestry maps each data object to a root node whose ID is equal or closest to the GUID of the object. When a node stores or retrieves a data object, it always sends the request to the root of the object. More specifically, when a node i publishes a data object v it stores, it sends the publishing message to the root of the object O. Each of the intermediate nodes on the path from i to the root node stores a location mapping (v, i). Upon receiving a request message for the object v, each node on the path to the root node checks if it has the location mapping for v. If it does have the mapping, it redirects the message to the node i who stores v; otherwise, it forwards the message to the next hop towards the root node. As a result, for any data object, the routing paths of the request messages form a unique spanning tree, with the root being the root of the object.

Figure 2.4 illustrates an example of publishing data objects in Tapestry. Both the node "4228" and the node "6712" store a data object O. Note that the node "6471" is the root for this data object. When the node "4228" and "6712" send publishing messages, each of the intermediate nodes along the paths towarding to the object's root creates a location mapping, i.e., a pointer for the object O to the publishers "4228" and "6712."

Figure 2.5 illustrates an example of how nodes retrieve a data object O. There are three nodes, "54FE," "7452" and "656C," retrieving the data object O by sending querying messages towards the root of this object. When the message from "54FE"

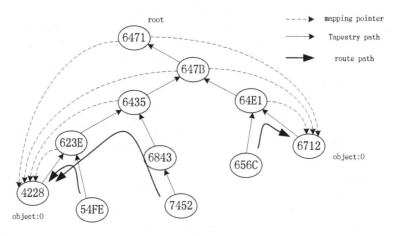

Fig. 2.5 An example of retrieving a data object in Tapestry

is received by "623E," the latter finds that it has maintained the location mapping for the object O (which points to "4228"), and thus forwards the querying message directly to the publisher "4228," rather than further forwards the message to the root "6471." Similarly, the messages from "7452" and "656C" are sent directly to "4228" and "6712" respectively. By doing so, Tapestry efficiently improves the routing throughput and latency.

When a node i is joining a Tapestry system, it is assumed that it knows (or can find) an active node which is already in the system. The data objects which should be rooted at node i must be migrated to i. During the migration, node i constructs its own routing table, and notifies other nodes so that they can insert i into their routing table.

When a node i is leaving the system, Tapestry solves the problem in two cases: voluntary node deletion and non-voluntary node deletion. In the first cast, the leaving node i notifies all nodes that are related to i and moves the objects it maintains to the new root. In the second case, node i leaves or fails without any notification. In this case, nodes rely on periodical messages to detect whether the outgoing link and node fails. Furthermore, Tapestry builds redundant pointers in routing tables to improve robustness to node departures. A combination of these techniques retains nearly 100 % success rate for routing messages.

2.7 Viceroy

Viceroy [46] is constructed on the butterfly network. In Viceroy each node has two associated values: *id* and *level*. *id* is the identity of the node in the network, which is a random value in the range $[0, 1)$ and is fixed throughout its participation; *level* is a positive integer in the range $[1, \log N]$ and changes as the network evolves.

Nodes in Viceroy form three types of sub-topologies: a *general ring*, which is similar to the ring in Chord and is constructed by the successor and predecessor relationships, multiple *level rings*, where all nodes in the same level form a ring, and a *butterfly* network where each node connects to proper nodes to form a butterfly network. In the butterfly network, each non-leaf node at level l connects to two nodes at the lower level $l + 1$; such links are referred to as the left and the right down links respectively. The left down link connects to the node which is the clockwise closest, and the right down link connects to the node at about $1/2^l$ away at level $l + 1$. Additionally, each node at level $l > 1$ also connects to one node at the upper level $l - 1$ which is numerically the closest node to the local node (such links are referred to as up links).

The routing model and the data model are unified via a key primitive, LOOKUP, in Viceroy. This primitive is used not only to maintain the network during nodes' joining and leaving, but also to data retrievals. The primitive LOOKUP consists of three phases. Firstly, the root node at level 1 is found through the up link. Secondly, the next-hop node is selected using the down links. When a node at level l calculates the distance between the destination and itself to determine which of the two down links (i.e., left or right) should be used to forward messages. More specifically, if it finds that the distance is at least $1/2^l$, then it forwards the lookup message through the right down link; otherwise, the left down link should be used. Thirdly, when a message reaches a node without down links or overshooting the destination, the message is forwarded along the level ring until the destination node is found.

When a node i is joining the Viceroy system, it should join each of the three types of sub-networks. First, node i finds out its predecessor and successor in the general ring based on its ID. It inserts itself into this ring and gets the key-value pairs between i and $predecessor(i)$ from its successor. This procedure operates similar to Chord. Second, node i selects its level by a level selection mechanism and joins the level ring. Last, node i joins the butterfly by choosing the two down links and one up link.

When a node is leaving the Viceroy system, it removes all of the outbound connections and notifies all neighboring nodes so that they can find a replacement. Moreover, the leaving node must transfer the resources it maintains to its successor.

2.8 Comparison

All DHT variants discussed in the chapter focus on how to efficiently manage a large number of nodes and data in a distributed system, and each variation has its unique characteristics.

In this section, we will discuss their similarities and differences from the perspectives of how the overlay network is constructed, the distance metric, the routing and data model, and how node dynamics are handled.

2.8.1   Overlay Network Topology

Chord is the simplest variation of DHT. It is a one-dimensional ring where all IDs
of both data and nodes are arranged clockwise in the ascending order. Every node
maintains the data whose IDs fall in the range bounded by the node's ID and its
predecessor's ID. CAN is featured with a d-dimensional hyper-cube structure. Its
d-dimensional Cartesian space is divided into a number of non-overlapping zones,
each of which is maintained by one node. Any data object is mapped into the d-
dimensional hyper-cube as a point in one of the zones. GISP is a structureless
variation of DHT, where there is no limitation for the number of connections among
nodes. Any two nodes have a direct connection if they know each other and they
are alive. Kademlia, Pastry and Tapestry all form a one-dimensional structure which
could be thought of as a tree structure. The nodes' identifiers constitute the leaf
nodes in the tree. They use fixed-length bit strings as the IDs for nodes and data
objects (Kademlia and Tapestry use 160-bit IDs; Pastry uses 128-bit node ID and the
object ID is at least 128 bits long). Additionally, Kademlia organizes the network
with the XOR operation, so its structure is a special tree referred to as the XOR-
tree [47]. Pastry uses a ring structure to assist the routing when the tree structure
can't find a proper target. Viceroy constructs a butterfly structure, which is the most
complex one in the aforementioned variations.

The levels of connectivity available in these variants also differ significantly.
More specifically, CAN has strong connectivity between nodes due to the d-
dimensional hyper-cube structure, and Viceroy benefits from the property of
butterfly structure, while in all other variants, nodes are less connected. Thus nodes
in these variants with less connectivity have to maintain multiple other nodes as their
neighbors (these nodes are not its numerical neighbors). For instance, each node
in Chord maintains a finger table, containing $O(\log N)$ nodes which are not the
numerical neighbors on the one-dimensional ring. The distances from this node to
the nodes in the finger table are half of ring perimeter, one-quarter of ring perimeter,
one-eighth of ring perimeter, ..., so on and so forth. As a comparison, Viceroy
maintains three types of connectivity (i.e., links on the general ring, a level ring
and a butterfly network) which allow nodes to connect not only to their numerical
neighbors but also to other nodes. Due to the butterfly structure, each node in
Viceroy only maintains $O(1)$ peers. The enhancement of the connectivity in these
structures significantly improves the efficiency of routing.

2.8.2   Distance

The distance metric, $d(i, j)$, measures the distance between two nodes (or data
objects) of i and j in a distributed system. This metric is the key and basis for
routing and data retrieval operations. As a result, the difference in the distance
metrics used by different variants leads to different routing strategies.

In Chord, the IDs of nodes and data objects are treated in the same manner. The distance from i to j is defined as $d(i, j) = (j - i) \mod 2^m$, where m is the number of bits in the key and node identifiers. In CAN, the distance is the Euclidean distance in the d-dimension space. From the geometric point of view, it is the distance between two points in the d-dimensional hyper-cube. In GISP, the distance between two data objects is the differences between their IDs. For the distance between two nodes, GISP introduces a new parameter called "peer strength," which represents the capability of a node. More specifically, the distance between two nodes with IDs i and j is $d(i, j)/(2^{s_i - 1} 2^{s_j - 1})$, where s_i and s_j are the "peer strength" values of these two nodes. This distance metric assigns a greater responsibility to the nodes with a more powerful capability. In Kademlia, the distance between node i and j is the bitwise exclusive OR (XOR), i.e., $d(i, j) = i \oplus j$. This metric has extremely low computational complexity than the Euclidean distances (e.g., adopted by Chord and CAN), and it does not need an additional algorithmic structure for discovering the target in the nodes which share the same prefix. In Pastry and Tapestry, the distances both assessed digit by digit as in Plaxton [48]. In Viceroy, the IDs of data objects and nodes are numbers in the range $[0, 1)$, thus the distance from node i to j is $d(i, j) = (j - i) \mod 1$, similar to the distance metric used in Chord.

The distance metrics of Chord, Kademlia and Viceroy are unidirectional, which means that for any given key k and a distance $D > 0$, only one key k' can satisfy $d(k, k') = D$. In other systems the number of nodes satisfying this condition is more than 1. This also means that Chord, Kademlia and Viceroy can determine a unique node by distance for the routing purpose, while all remaining variants need to determine which node is the next hop by additional parameters. Besides, The distance metrics used in Chord and Viceroy are directed, which means that $d(i, j) \neq d(j, i)$ in most situations.

2.8.3 Routing and Data Model

We compare the routing and data models adopted by different variants from three perspectives: the routing table, the routing path selection and length. Below we assume that there are N nodes in the network.

Routing Table

Each node in all DHT variants maintains the information of a set of other nodes for the purpose of routing. However, different variants maintain different types of nodes.

In Chord, each node stores three lists of nodes: a predecessor node, a finger table consisting of up to m entries/nodes (note that the size of the ID space is 2^m), and a successor node list with r entries. The total number of nodes maintained by each node is $O(\log N)$. In CAN, each node in the d-dimensional hyper-cube maintains a list of $2d$ neighbors, which is independent of the total number of nodes in the system (i.e., N). This property is desirable since nodes keep a constant number of neighbors regardless of the system size. In Kademlia, the length of node IDs is 160 bits. For any i ($0 \leq i < 160$), each node keeps a k-bucket, in which at most k nodes are stored. The distance from any of the nodes in the ith k-bucket to the local node is in the range $[2^i, 2^{i+1}]$. Therefore, each node in Kademlia maintains at most $160 \times k$ neighboring nodes, which is equal to $O(\log N)$ in essence. In GISP, each node maintains as many nodes as possible, but GISP dose not explicitly define strategies for managing such information. In Pastry, each node maintains three sets of nodes: a routing table consisting of approximately $\lceil \log_{2^b} N \rceil \times (2^b - 1)$ nodes (recall that the levels are denoted by b bits), a neighborhood set of M nodes, and a namespace set of L nodes. The typical values for L and M are 2^b and 2×2^b, respectively. In Tapestry, each node has a location mapping which consists of a neighbor map with $\log_{2^b} N$ levels. To some extent, this is similar to the routing table in Pastry. Each level contains $c \times 2^b$ nodes, where c is chosen for redundancy of routing nodes. The number of nodes every node maintains in these two variations is $O(\log_{2^b} N)$. Further more, Tapestry nodes also maintain pointers to the nodes who publishes the objects to the local nodes, which reduces the lookup time greatly. In Viceroy, each node maintains a constant (small) number of nodes. The number of nodes a node maintains is at most 7.

In CAN and Viceroy, each node maintains only a constant number of nodes, regardless the size of the network. On the contrary, in most of other variants, each node maintains $O(\log N)$ nodes, which is a varying number dependent on the size of the network. In summary, Viceroy nodes maintain the least number of nodes [i.e., $O(1)$], while others maintain $O(\log_2 N)$. Thus Viceroy costs least to maintain the complete structure; however, the structure of Viceroy is more sensitive to node failures than other variants.

Path Length

The average length of routing paths is an important parameter that suggests the efficiency of routing: the less number of hops a routing path has, the more efficient the routing is.

In Chord, the length of the routing paths is no more than $O(\log N)$ hops, as a result of the finger table which allows to search in an exponential manner. In CAN, due to the d-dimensional hyper-cube structure, the average routing path length is $\frac{d}{4} N^{\frac{1}{d}}$ and any node can be reached by another node in $O(dN^{\frac{1}{d}})$ hops. In Kademlia, Pastry and Tapestry, the average length of routing paths can be improved to $O(\log_{2^b} N)$, where b bits are merged to denote a digit. For example, $b = 4$ means that the IDs are hexadecimal and the 160-bits ID is regarded as a 20-digit hexadecimal number. when $b = 1$, Kademlia, Pastry and Tapestry are the same as Chord in terms of the efficiency of routing. In Viceroy, the average length of routing paths is also $O(\log N)$. In GISP, since any two nodes may be connected, it is challenging to analyze its average routing path length of GISP.

Path Selection

Each node in the above DHT variants knows only partial information about the network. However, DHT ensures efficient routing path selection with such partial information.

More specifically, each node tries its best to find out the node closest to the destination. Here the "closest" node is chosen based on nodes' numerical distances to the destination. The distance metrics used in different variants are different; as a result, the paths from the same source node to the destination are different in different variants. All DHT variants adopt a locally optimal solution in each step, which may achieve the global optimal solution. This is the reason why DHT-based overlay networks perform efficiently in large-scale distributed systems.

2.8.4 Node's Joining and Leaving

When a node i wants to join the system, it is assumed that it knows at least one active peer j which is already in the system. In Chord, the active node j helps node i to find its successor. Node i joining the system requires that some nodes in the system should update their finger tables. With a stabilization protocol running periodically, nodes update their successor nodes and their finger tables gradually, which costs $O(\log N)$ times and $O(\log_2 N)$ messages. In CAN the active node j introduces node i into the system which allocates a zone for i to maintain. Since all nodes only maintain a list of neighboring nodes, when a node joins in CAN, the neighbors of this new node should update their neighbor lists. This is sufficient to generate the correct routing path without a stabilization protocol. Thus the joining operation in CAN costs $O(1)$ times. In Kademlia, the joining node i starts a LOOKUP operation to insert itself into other nodes' k-buckets and constructs its own k-buckets gradually. In GISP, the network needs extra time to stabilize when a node comes. In Pastry and Tapestry, node i must try its best to learn enough number of peers and tells others to memorize it. In Viceroy, node i constructs the three types of links by the JOIN operation [46].

When a node i leaves the network, affected nodes must update the node information they maintains. Chord runs a stabilization protocol to update the nodes' successors and finger tables. In CAN, each normal node sends update massages to its neighbors periodically. When neighbors find a node has left, one of them takes over the failed node's zone and notifies other neighbors. In GISP, when a node finds that some nodes are unreachable, it deletes these nodes from its routing list. In Kademlia and Pastry, there are no specific protocol to handle node departures; rather, they detect the target node before routing. This method reduces the impact of node leaving, but it may increase the latency of message routing. In Tapestry, nodes use heartbeat messages to detect whether a node is alive. In Viceroy, when a node leaves voluntarily, the leaving node should execute the LEAVE operation [46] before its departure. Otherwise, when node failures occur, the structure of Viceroy should be repaired by executing the LOOKUP operation (Table 2.1).

Table 2.1 Comparison of classical DHT variants. (a) Comparison from topology, distance and path length (b) Comparison from path selection, number of peers, node's joining & leaving

(a)

Algorithm	Overlay network topology	Distance from i to j	Routing path length
Chord	One-dimensional ring	$(j - i) \mod 2^m$	$O(\log N)$
CAN	d-dimensional cube	Euclidean distance	$\frac{d}{4} N^{\frac{1}{d}}$
GISP	Structureless	Objects: the difference of the two IDs; Nodes: $(i, j)/(2^{s_i-1} 2^{s_j-1})$; s_i, s_j are "peer strength"	Uncertain
Kademlia	XOR-tree	$i \oplus j$	$O(\log_{2^b} N)$
Pastry	Tree+ring	Assessed digit by digit	$O(\log_{2^b} N)$
Tapestry	Tree	Same as Pastry	$O(\log_{2^b} N)$
Viceroy	Butterfly	$(j - i) \mod 1$	$O(\log N)$

(b)

Algorithm	Routing path selection	# of maintained peers	Node's joining	Node's leaving
Chord		$O(\log_2 N)$ construct	Find successor; protocol	Run stabilization
CAN		$2d$	Generate neighbor list	Update neighbor lists
GISP	Greedy algorithm	As many as possible	Generate routing list	Delete failing nodes from routing list
Kademlia		At most $160 \times k(O(\log_2 N))$	Constructs k-buckets	Detect the target node before routing
Pastry		$O(\log_{2^b} N)$	Generate routing table, neighborhood set and a namespace set	Detect the target node before routing
Tapestry		$O(\log_{2^b} N)$	Construct the routing table	Heartbreak message
Viceroy		At most 7	Construct the three kinds of links	Repaired by the LOOKUP operation

Chapter 3
DHT Platforms

Based on the theory of DHT, many researchers develop platforms, that implement different kinds of DHT and provide interfaces and services to applications. DHT translation from theory to platform is not a simple work. In this procedure many problems and requirements will be exposed, such as load balance, multiple replicas, consistency, latency and so on. Some platforms only complete the basic functions including implementing specific DHT and providing interfaces to the upper applications, such as Open Chord, Chimera, FreePastry and Khashmir. Besides the fundamental implementation, some also supply specific services, such as CoDeeN and CoralCDN for caching, hazelcast for data distribution. Some supply additional guarantee, such as GNUnet for privacy and security. Some focus on providing a platform for connecting all kinds of devices, such as JXTA. In the following, we introduce the key design properties and characteristics of several DHT implementations in both academic/open source platforms and commercial platforms respectively.

3.1 Academic and Open-Source Platforms

In this section, we introduce 11 platforms that implement various DHT techniques, including Bamboo, CoDeeN, CoralCDN, OpenDHT, JXTA, GNUnet, Open Chord, hazelcast, i3, Overlay Weaver, Cassandra. They are all open-source platform, that are free and allow other people to contribute to the system, which facilitates the platform to grow and improve, but on the other hand, limits the maintenance and stability of the system.

H. Zhang et al., *Distributed Hash Table; Theory, Platforms and Applications*,
SpringerBriefs in Computer Science, DOI 10.1007/978-1-4614-9008-1_3,
© The Author(s) 2013

3.1.1   Bamboo

Bamboo [11] is a Pastry based DHT platform written in Java. The first release was written in December 2003. The latest version was published on March 3, 2006. Although Bamboo is built on the base of Pastry, it has several improvements especially in peers' joining and leaving problem, which is called "churn."

Bamboo performs well under high levels of churn [20] by three technologies. The first one is "static resilience to failures." This technology allows node to route messages even before recovery. The second one is "timely, accurate failure detection." In a P2P system where churn happens normally, it is quite often to send messages to a node that has left the system. So the timely failure detection is very important. Bamboo detects the failure by setting a proper timeout. If the timeout is too short, many massages would be re-routed and the target node is mistaken as a failed node. On the contrary, if the timeout is too long, the requesting node would waste time waiting the response from a left node. For choosing an accurate timeout, the nodes in Bamboo probe the latency actively. Furthermore, Bamboo uses recursive routing to solve the problem that how to actively probe any node in the network. The third one is "congestion-aware recovery." In this technology, Bamboo simplifies the Pastry's joining algorithm, which allows a node to join the network more quickly. In the leaf maintenance, Bamboo adds a "pull" operation which is a reply to the "push" message of nodes, while Pastry only has a "push" operation, that a node sends its entire leaf set to some neighbor nodes which is randomly chosen in the set. This "feedback" method greatly increases the consistency of the information that nodes maintain, especially in high churn situation. Bamboo provides two algorithms called "global tuning" and "local tuning" respectively, which optimize the route table all the time. Therefore, you may find that the nodes in Bamboo keep changing their neighbors, even no node joins or leaves the network.

Since Bamboo is constructed by Berkeley, it incorporates into other Berkeley projects easily, such as OceanStore and PIER [49]. Meanwhile, Bamboo is running as an OpenDHT project [50], which allows anyone to put and get key-value pairs into it only using XML RPC or Sun RPC. This make it much easier to run the DHT individually.

3.1.2   CoDeeN

CoDeeN [14] is an academic content distribution network (CDN) for PlanetLab [51] by the Network Systems Group at Princeton University. It consists of high-performance proxy servers which are deployed on PlanetLab nodes. Users are provided a fast and robust web content delivery service. CoDeeN has been built since 2003. At present it is still under development and tries its best to provide continual service.

CoDeeN reduces the response time of Internet by caching the web pages and resources of the remote sites on the proxy servers. These servers locate all over the world, so users can get the resources from the nearest proxy server instead of the original sites. If someone wants to use CoDeeN, she should pick the nearest proxy server from the CoDeeN proxy server list. Meanwhile, this cache technology also reduces the burden of the web site so that can support more users.

CoDeeN also does some work in privacy and security. All accesses via CoDeeN are logged and the logs are monitored for the abuse. CoDeeN uses semi-restricted proxies with several protections against the abuse. For instance, users cannot access the sites containing licensed contents. CoDeeN protects the resources with IP-address restrictions. A number of known virus and attack signatures are tested so that CoDeeN could ban clients attempting to use the attacks.

CoDeeN is an implementation of DHT, which focuses on the web caching. Although it is an academic platform, it provides a large amount of information for the related research, which improves the performance of commercial CDN.

3.1.3 CoralCDN

CoralCDN [7] is a peer-to-peer content distribution network, which is comprised of world-wide network of web proxies and nameservers. The first version of CoralCDN online was deployed in March 2004, which also runs on the PlanetLab [51]. Unlike CoDeeN that users configure the proxy server manually, CoralCDN works in another way. If Users want to access http://www.xxx.com/, they only add a suffix in the way of http://www.xxx.com.nyud.net:8090. CoralCDN chooses the optimal proxy server automatically.

Figure 3.1 shows the process of the CoralCDN. At first, a client sends a request for http://www.xxx.com.nyud.net to its local resolver. Upon reviving the request, the resolver transmits it to a Coral DNS server. The DNS server probes the round-trip time to the client. According to the result, the DNS server finds out the nearest http proxy server to the client and returns it to the client. Then the client sends HTTP request http://www.xxx.com.nyud.net:8090 to this specified proxy. If the proxy caches the proper object, it will return the web pages to the client. Otherwise, the proxy will look up the object on other proxy in Coral. Only if there is no proper resource in Coral, the original website will be accessed.

CoralCDN is built on a key-value indexing infrastructure Coral, which is based on distributed sloppy hash table (DSHT) that differs from traditional DHT in several aspects [52]. In DSHT, the PUT operation which puts the key-value pairs in the network does not only consider the distance but also avoids hot spots. Thus the resources would be stored far from the positions of their IDs. In the routing procedure, the router determines the next target which is not the node closest to the destination, but the one whose distance to the destination is closest to half of the local router's distance.

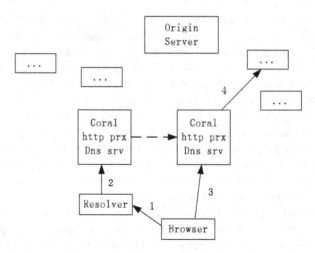

Fig. 3.1 The process of CoralCDN

Coral builds a XOR-tree structure like Kademlia and introduces locality information. Coral uses several levels of DSHT to build a hierarchical routing structure, which is divided by the delay. The higher the level is, the smaller the delay is and the fewer the nodes are in this level. If a request comes, a node in Coral will search proper resources on the highest level because of the smallest delay. If the resources are not found, then the lower level is used. Delay implies the geographical distance of two nodes. So the nodes at higher level probably locate in a smaller geographical coverage.

CoralCDN introduces several properties to extend the classical DHT, especially for avoiding hot spots and increasing speed. In the past years CoralCDN worked well. M. J. Freedman [53] had observed it for a long time, which helps researchers to learn CoralCDN. Compared with CoDeeN, CoralCDN is an open-source implementation, which allows more operability to improve and contribute to it.

3.1.4   OpenDHT

OpenDHT [50,54] is a platform which provides free and public DHT services. It was established in 2005. Unlike usual DHT model, users of OpenDHT do not need to run as a DHT node. The OpenDHT infrastructure runs the OpenDHT code and provides interfaces to users to connect to it. User nodes issue RPC (Remote Procedure Call) to upper applications. OpenDHT nodes act as gateways accepting the RPC from clients.

OpenDHT places a high premium on simplicity. There are three interfaces to simplify the operations and usages for users. In the three interfaces, routing model is the most general one. It provides general access to each of the nodes on the routing path for the input key. It allows users to invoke arbitrary code at the nodes on the routing path. Lookup model is another interface which is somewhat less general. Compare with routing model, lookup model only provides general access to the destination node for the input key. Correspondingly, code invocation is only allowed on the endpoint. These two interfaces allow application-specific code, which is the true power of the two interfaces. Based on the basic interfaces, developers can generate applications and services with abundant functionality. The third is storage model, which is far less flexible. It provides the $put(key, value)$ and $get(key)$ operations to support storage. This interface does not allow invoking application-specific code, which limits the applications it supports, but it also makes the interface simplicity to use and support.

OpenDHT agrees that public facilities provides limited storage. So the data in OpenDHT has limited lifetime. Moveover, OpenDHT protects the storage from some attacks. It presents two ways, which are *immutable puts* and *signed puts*, to fight against malicious attacks to data. Among other things, OpenDHT provides REMOVE function which is handled like regular insertion of keys.

3.1.5 JXTA

JXTA [55] is an open-source P2P platform started by Sun Microsystems in 2001. It consists of several protocols that enable each connected device on a network to exchange messages independently of the underlying network topology. The goal of JXTA is providing services and infrastructure for P2P applications which are independent from operating system and language.

JXTA has three distinct layers [56]. The lowest layer is platform, which contains the core and fundamental functionality, such as peer pipes, peer monitoring, peer groups, etc. The upper one is service, which provides several access to the JXTA protocols such as searching, file sharing, indexing and so on. Application is the top layer, which accesses the JXTA network and utilities based on services. XML documents are widely used to describe services and information available on the JXTA network because of their popularity and ability to be read easily by many languages.

There are six protocols which construct the core functionality and standard services [57], including:

- Peer resolver protocol (PRP) which is a basic communication protocol providing query/reponse services.
- Endpoint routing protocol (ERP) which provides a set of query messages that helps peer route messages to the destination.
- Peer discovery protocol (PDP) which helps peers to announce advertisements and to discover other peers, groups and other information.
- Rendezvous protocol (RVP) provides mechanisms that enable propagation of messages within a peer group.

- Peer information protocol (PIP) which is used to query status information form peers.
- Pipe binding protocol (PBP) which builds a pipe or interface between peers for communicating and routing.

These protocols provide basic functions in P2P network computing. They hide many details in the lower level. Thus it is much easier to develop applications on JXTA. Nowadays many applications and services are developed on JXTA.

The routing tragedy of JXTA is a loosely-consistent DHT walker which combines DHT index and a limited range walker [58]. In this way JXTA can work well both in the high-churn-rate situation and the steady network environment.

3.1.6   GNUnet

GNUnet [15,59] is a framework for secure P2P networking. Anonymous distributed file sharing based on reputation is the first service implemented on the networking layer. GNUnet is written in C language and currently runs on Windows™, Linux, Mac OS™, BSD and Solaris™. GNUnet uses Kademlia style routing [60], which is popular in the implementations of DHT.

The goals of GNUnet are deniability for all participants, distribution of contents and loads, and efficiency in terms of space and bandwidth [61]. It provides authentication and protective security against particular attacks in the network layer. Meanwhile, a file sharing service providing full anonymity is implemented in the application layer. Every peer in GNUnet inserts content anonymously and claims ignorance. Content migration is used to prevent the publisher being located. Hence, the adversary cannot confirm the content publisher unless performing full traffic. The traffic is protected by encryption and encoding, which achieve the aim that none of the intermediaries knows the content while the receiver can decrypt it.

In GNUnet, the file is split into several GBlocks. Each block is only 1k. So it is convenient to migrate file even the file is large. The file may be maintained by multiple nodes. It avoids the traffic-burst in migration since the load is distributed to plenty of nodes. The blocks are encrypted, so the content is hidden from the intermediaries in transmission, even the maintainers if they do not have keys.

GNUnet is able to avoid content from malicious hosts. GNUnet uses a double hash method for content and query. The first hash value is used as the encryption key. The second hash value (the hash value of the first one) is used to locate the data. Since the data transmitted in the network is encrypted, the privacy is preserved from malicious nodes. Moreover, using hash value as the key solves another challenge. If two parties insert the same file in the network independently, they will use the same key (hash values of files are the same). The two versions can replace each other even they are encrypted.

GNUnet applies indirection mechanism for anonymity. Indirection hides the source, since it claims that it just indirects the packets. However, this scheme costs too much if all the nodes in the transmission path. In GNUnet, indirect queries are decided by the receivers freely or randomly whether or not to indirect the reply, which decreases the cost without reducing the security. Furthermore, GNUnet reduces the influence of malicious nodes as possible as it can. When a new node joins GNUnet, it is treated as untrusted one that the established nodes reply the new node's query only if they have excess bandwidth. The reputation tragedy is used so that the malicious nodes have little influence. In this way, GNUnet reduces the harm of malicious nodes greatly.

The significant distinction between GNUnet and other file sharing system like Gnutella and Napster is that GNUnet gives stronger security guarantees that it is more resistant to attacks. In 2011, GNUnet was again reachable via IPv6. Now it fully supports IPv6.

3.1.7 Open Chord

Open Chord [62] is an implementation of the Chord DHT. It provides an interface for P2P applications for quickly storing and retrieving data from Chord. Open Chord is written by Java and is distributed under GNU General Public License (GPL), which allows Open Chord to be used and extended for free.

Open Chord is divided into three layers [63]. The lowest layer is communication layer that employs communication protocol. The protocol is based on a network communication protocol such as Java Sockets. On the communication layer, the communication abstraction layer is implemented. This layer hides the details of the communication and provides interfaces for synchronous communication between peers. On top of the communication abstraction layer, a Chord logic network resides. This layer provides two interfaces to applications for storing, retrieving and removing data in the Chord DHT synchronously and asynchronously. This layer implements the properties of Chord DHT described in [6].

Open Chord provides interfaces and APIs so that applications can be easily implemented when they want to employ Chord DHT. Now the latest version is Open Chord version 1.0.5, but the manual is still for Open Chord version 1.0.4. This manual shows some limitations of version 1.0.4, such as prohibition of remote class loading, changing the communication protocol easily, trust in all the participant of the DHT and so on. Nevertheless, Open Chord still is a good implementation of Chord.

3.1.8 Hazelcast

Hazelcast [8] is a clustering and highly scalable data distribution platform for Java, from which developers can easily design and develop highly scalable and reliable applications for their businesses.

In Hazelcast the data is almost evenly distributed across all nodes, which means each node carries ($\frac{1}{n}$ "total data") + "backups," where n is the number of nodes in the cluster. If a node fails, the data it holds will be dynamically redistributed to the remaining live nodes by the backup replicas stored on other members. When a new node joins, it will be responsible for almost ($\frac{1}{n}$ "total data") + "backups" to reduce the load on others. In Hazelcast's data structure all nodes have equal rights and responsibilities. The distribution of data and organization of the nodes are based on DHT. So there is no single cluster master that may cause single point failure.

Hazelcast provides many distributed technologies to support the distributed data management [64]. It implements the *Distributed Queue, Distributed Set, Distributed List* and *Distributed Map* based on `java.util.{Queue, Set, List, Map}`. A distribution mechanism *Distributed Topic* is provided for publishing messages that are delivered to multiple subscribers. Furthermore, *Distributed Lock* is deployed and *Distributed Events* is used to satisfy the distributed environment. Moreover, Hazelcast has other functionality and features which make Hazelcast manage the distributed data better.

Although Hazelcast is developed by a company, it is a real open source platform, where the source code can be downloaded at http://code.google.com/p/hazelcast/. ZooKeeper [65] is another system which provides similar function with Hazelcast, but instead of Hazelcast's DHT, ZooKeeper is based on master/slaver model.

3.1.9 i3

i3 [66, 67] is an overlay-based Internet Indirection Infrastructure that offers a rendezvous-based communication abstraction to provide services like multicast, anycast and mobility. Every device connecting to i3 infrastructure is associated with an identifier, which could be used to obtain delivery of the packets. i3 stores the triggers. When one host wants to send packet to another, the packet will be assigned an identifier too. i3 would transmit the packet to the proper destination based on the trigger of the identifier. For example host R inserts a trigger (id, R) into i3, then all the packets with identifier id would be received by R.

Multicast, anycast and mobility are three fundamental communication services that i3 provides. Suppose there is a mobile host with a trigger (id, R) in i3. When the host moves from one location to another, it will be assigned a new address R'. The trigger in i3 is changed from (id, R) to (id, R'). So the sender need not to be aware of the current address or location of the mobile host, instead he only knows the identifier of the destination. If some packets will be send to a group of hosts, all the members of the group register triggers with the same identifier. What's more, i3 provides another scheme to support anycast. In this case, the identifiers of receivers share a common prefix p. When the packet with the identifier $p|a$, where p is the prefix and a is the suffix, is sent to the group, the corresponding receiver would be chosen by the longest matching prefix rule. Furthermore, i3 supports the

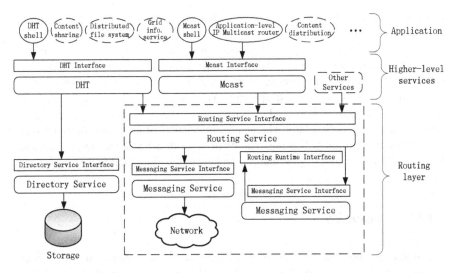

Fig. 3.2 Decomposition of runtime in Overlay Weaver

mechanism of identifier stack, which allows applications to insert some transcoders in the routing path of packets. This service supported by i3 greatly facilitates many applications such as watermarking in the video stream, video format conversion.

i3 is an open source implementation of Chord. An instantiation of i3 has been running on the Planetlab. i3 can be used in many useful applications, such as connecting to home machines, secure intranet access, middle-box applications. i3 works like a cloud, in which the id-address pairs are stored. The hosts do not need to take care of the accurate address of the destination. They only assign the identifier of target to the packets which will be sent, regardless of multicast or unicast, mobile host or fixed one. i3 hides the details of the implementation to the applications, which gives us an easy way to use it.

3.1.10   Overlay Weaver

Overlay Weaver [68,69] is an overlay construction toolkit which provides a common API for higher-level services such as DHT and multicast. Overlay Weaver is an open source implementation. It currently implements many structured routing algorithms including Chord [6], Kademlia [33], Koorde [24], Pastry [43] and Tapestry [44], and also supports unstructured algorithms.

Overlay Weaver is designed in a modular manner, i.e., it can be divided into several components. Figure 3.2 illustrates the components of Overlay Weaver. The lowest layer is routing layer, which is corresponding to key-based routing. This layer has been decomposed into three parts: Messaging Service, Routing Algorithm and Routing Driver. Messaging Service deals with communication between hosts. Routing Algorithm implements many kinds of routing algorithms. Routing Driver conducts the common routing process and provides uniform interface to the upper layer. Based on the routing layer, the higher-level services layer is deployed. In addition to DHT, it provides the Mcast, which performs a multicast on an overlay. This level supports abstract interface to applications layer, which makes it easy to develop applications on Overlay Weaver. Furthermore, Overlay Weaver also provides Distributed Environment Emulator which can emulate tens of thousands of nodes on a single computer virtually. It brings results of algorithm researches to applications directly, which gives developers and researchers much more facilitation.

3.1.11 Cassandra

Cassandra [9], originally designed by Facebook^{TM}, is a distributed structured storage system deployed on a large amount of commodity servers. It provides a NoSQL database to precess plenty of data and has been employed by many famous companies like Facebook^{TM}, Twitter^{TM}, Digg^{TM}, Cisco^{TM}, etc. Cassandra is an open source implementation since 2008. Now it is developed as an Apache [70] top level project.

Cassandra combines the data model of Google^{TM}'s BigTable [71] and distributed architecture of Amazon^{TM}'s Dynamo^{TM} [12]. It is designed to run on large-scale distributed systems handling very high write throughput and achieving scalability and availability. Rather than exception it treats failures as the normal situation. It provides database interface composed of three APIs: insert(*table; key; rowMutation*),get(*table; key; columnName*), delete(*table; key; columnName*). It is so easy to use, but on the contrary the internal implementation of Cassandra is not an easy job. Cassandra is hoped to process the ability to scale incrementally, so consistent hashing using order preserving hashing is used. Meanwhile, a token method is used for scaling the cluster. High availability and durability are achieved by replication. Cassandra provides several replication policies to meet various situation and requirements. For failure detection, a modified version of the Φ Accrual Failure Detection [72] is used, which introduces a dynamically adjusted parameter Φ which reflects network and load conditions.

Cassandra can be treated as a NoSQL database with four prominent characteristics [73]. Firstly, Cassandra is decentralized, where no master exists which would introduce single point of failure. Secondly, read and write throughput increase linearly with addition of new nodes. The performance improvement can be realized without downtime and interruption to applications. Thirdly, multiple replicas of data in Cassandra exist at any time for fault-tolerant. Node failure can be solved without downtime. Lastly, Cassandra supports a tunable level of consistency, which allows users to choose a tradeoff between write and read in different application scenarios. Cassandra is an outstanding distributed database that achieves high update throughput with low latency, even it still is perfected by Apache committers and is contributed by many corporations [74].

3.2 Commercial Platforms

In this section, we present four commercial DHT platforms, including WebSphere eXtreme Scale™, Dynamo™, SandStone™, and Service Routing Layer.

3.2.1 WebSphere eXtreme Scale

WebSphere eXtreme Scale™ [10] is a proprietary DHT implementation by IBM™ used for object caching. It performs as a powerful distributed cache system to speed the access to data.

For improving the performance of computer, engineers add several levels of cache between memory and CPU. The cache can increase the speed of access data. This technology can be used in network similarly. WXS™ puts two levels of cache between data source and users that greatly speeds the access. However, the key difference between computer and network in caching is that the caching mechanism in computer is centralized while it is distributed in network. WXS™ organizes the cache in DHT way. In WXS™ the most fundamental component is the grid. Data is stored as key-value pairs in the maps. Several maps are contained in the grid. The map sets can be partitioned into parts which are maintained by multiple containers. In this way the cached data are organized between plenty of machines.

Figure 3.3 illustrates the outline of high performance access using WXS™, where the client is searching proper data. In this procedure, ObjectGrid API firstly searches the data in the Near cache. If nothing is found in the Near cache, it will locate the shard in the grid who contains the querying data. Here shard is the instance of data (or a portion of data split by WXS™). If the result is still not found, the data would be loaded from the Back-end datasource. In this scenario the client, Near cache, shard and datasource resembles CPU, L1 Cache, L2 Cache and main memory in computer respectively.

WXS™ provides distributed object caching essential for elastic scalability. Every map set has four or three partitions and each partition has one or two replicas [75]. The Characteristics of WXS™ are high performance and elastic scalability, but the performances are greatly influenced by the cache hit rate. So WXS™ should be deployed with adequate consideration so that the power of WXS™ would be leveraged sufficiently.

3.2.2 Dynamo

Amazon™ is one of the largest e-commerce operations maintaining a large number of servers and data centers all over the world. It runs a world-wide e-commerce platform to millions customers, which has strict requirements to the storage system

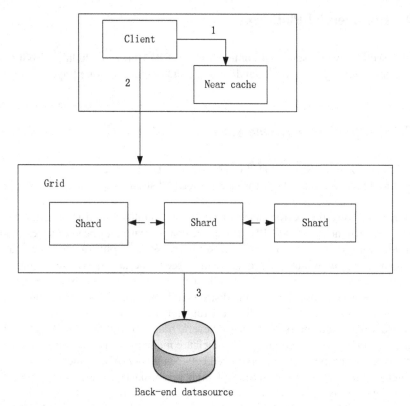

Fig. 3.3 Structure of ObjectGrid of WXS^{TM}

in terms of performance, reliability, efficiency and scalability. Dynamo^{TM} [12], a component of the platform of one of the largest e-commerce corporations which maintains a large number of servers and data centers all over the world, is a key-value completely distributed storage system that provides an "always-on" storage experience, satisfying the customer's strict requirements to the storage system in terms of performance, reliability, efficiency and scalability to Amazon^{TM}'s core applications and services. It provides a DHT-typical simple primary-key only interface to meet many services on Amazon^{TM}'s platform.

Dynamo^{TM} deploys many technologies to solve the problems in large-scale distributed storage systems and achieve high usability and performance. Firstly, Dynamo^{TM} uses a variant of consistent hash that each node is assigned to multiple points in the ring, which could balance the load by the number of virtual nodes that one physical node is responsible. Meanwhile, one piece of data is maintained by multiple nodes in the form of replicas, which could achieve high availability and durability. Thirdly, vector clocks are used to provide eventual consistency, and consistency of read operation is guaranteed by quorum scheme which requires $R + W > N$, where N, R, W are the number of replicas, the minimum number

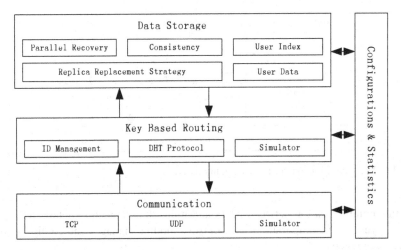

Fig. 3.4 The architecture of SandStone^{TM}

of nodes that must participate in a successful read or write operation respectively. Fourthly, hinted handoff and replica synchronization in which inconsistencies are detected by Merkle trees are used to handle churn and failures of nodes. In this manner, Dynamo^{TM} is running to support thousands of millions of users and many demanding services.

During these years, some criticisms about Dynamo^{TM} exist, especially Dynamo^{TM} is regarded as "a flawed architecture" [76]. However, Dynamo^{TM} is still running well and provides several core services. Argument and debate are not always bad things. On the contrary, intense discussion makes people know concept better, which helps us to perfect applications and services.

3.2.3 SandStone

SandStone^{TM} [13] is a DHT based key-value storage system developed by Huawei^{TM}. The name "SandStone" means that "the enormous sand-like tiny PCs' capability is united as an invincible cornerstone of distributed infrastructure." SandStone^{TM} is a highly decentralized, loosely coupled architecture with carrier grade performance. It performs good scalability, strong consistency, high reliability which can meet or exceed "five nine" high availability standards.

SandStone^{TM} architecture is depicted in Fig. 3.4. It consists of four components. The top layer is the data storage module, which takes charge of storage. It is composed of many functions and submodules to manage the data including data consistency verification, data storage and data restoring. The middle layer is the key based routing module, including ID allocation, DHT routing protocol and peer failure detection. The bottom layer is the communication module, which completes

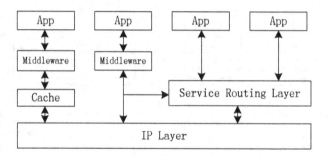

Fig. 3.5 The service routing layer's position in network

the intercommunication between peers and hides the connection details from the upper layers. The configurations and statistics module manages and configures the other three layers.

SandStone™ applies many technologies to increase the performance of the system. SandStone™ uses an ID allocation method called "Strip Segmentation" to let peers carry region indication. To makes good use of region information, SandStone™ is composed of a global DHT and several logical region DHTs, which are used not only to decrease the backbone network overhead, but also to reduce the latency. Furthermore, besides the finger table in Chord, SandStone™ makes a further step to add an extra One-hop Routing Table, which could reduce the latency further to meet the latency sensitive applications. To achieve high availability and durability, SandStone™ replicates data on multiple peers. Unlike Dynamo™ storing replicas successively, SandStone™ stores them more dispersedly. In order to keep these replicas consistent, SandStone™ also modifies the "quorum" technique to provide optimized eventual consistency.

3.2.4 An Overlay Platform: Service Routing Layer

In the past the network service providers usually offer IP packet delivery service and other IP-based services. With the growing requirement to multiple multi-level services, a *service routing layer* is proposed. It is an overlay platform [77]. Figure 3.5 shows the service routing layer lying on the IP layer. Some edge-based application and innovation can continue as before. Moreover, enhanced functionality and services can be offered by the overlay service routing layer. The overlay is programmable so that new functionality can be produced by combining a small set of basic defined functions. The service routing layer facilitates the deployment of new services and applications provided either service providers or the third parties, which is the goal of its implementation.

This platform by Cisco™ pays more attention to the combination of overlay layer with physical network. It is designed as a provider-hosted overlay in order to provide both overlay based applications and the ones IP layer directly supported. The routing layer provides a "topological proximity" function to employ some network layer topology information to the overlay layer. Furthermore, the platform takes the advantage that service provider has more ability to control the properties of network into account, so it also makes use of Diffserv to improve the QoS. The platform implements the relatively complex schemes in the above, and provides a set of simple functions to the applications, which are distributed hash table, proximity, caching, streaming support and events. Based on these functions, block functions can be extended and composed, which could implement a rich set of applications and features.

In this service provider-hosted platform employs services in DHT way, which gives services good scalability. A service descriptor mechanism is used to release services. The descriptors which delegate services are stored and located in the DHT. Services can be replicated on many places. With this scheme a service shows great scalability, and the allocated resources of it can be changed dynamically according to the popularity of the service.

3.3 Conclusion

Recently, many platforms of DHT have been implemented. All of these implementations provide interfaces to other applications or can be combined with other projects. In this chapter, we introduce some of the DHT platforms. Each of them has its unique features and characteristics. These platforms can be classified into two sets. One focuses on supplying some service itself and can be integrated with other applications and projects to supply richer services. Such as Bamboo, CoDeeN and CoralCDN that support caching service, GNUnet for anonymous distributed file sharing, hazelcast for data distribution, Dynamo^{TM} for object caching, Cassandra, Dynamo^{TM} and SandStone^{TM} for data storing. They improve the DHT for their features like Bamboo for high churn, GNUnet enhancing the security in communication. These platforms focus on some areas of DHT applications, but can be infrastructures integrated with other projects to extend services. The other set emphasizes on providing interfaces to other applications, such as OpenDHT, JXTA, Open Chord, i3, Overlay Weaver, Service Routing Layer. They do not care about supplying particular application. The goal of them are providing basic functionalities to upper applications. The platforms in former set have more independence that they can provide services without additional components. The platforms in latter set are more general that they focus on the basic functions for applications and services developed on them. The characteristics of each platforms are summarized at Table 3.1.

In the past several years, plenty of platforms of DHT are implemented. Besides the implementation above, there are still many platforms, such as SharkyPy [78] which is an implementation of Kademlia DHT, Chimera [79] which is a light-weight DHT implementation providing similar functionality as Tapestry and Pastry,

Table 3.1 Comparison of DHT platforms

Platform	Type	Properties
Bamboo	Open source	Improving Pasrty; performing well under high levels of churn
CoDeeN	Academic	Focusing on the web caching; providing strong privacy and security
CoealCDN	Open source	A content distribution network which provides an intelligentized interface to be used easily; using hierarchical DSHT which performs better than the normal DHT at avoiding hot spots and increasing speed
GNUnet	Open source	Using Kademlia style routing; providing strong security
Hazelcast	Open source	A clustering and highly scalable data distribution platform; good load balance and backup of distributed data
Cassandra	Open source	Storage system deployed on plenty of servers; NoSQL database;
WXS	Commercial	Distributed object caching; elastic scalability and high performance
Dynamo	Commercial	Distributed storage system; suited for large-scale application; high usability and performance
SandStore	Commercial	Storage system with carrier grade performance; improving the performance with many technologies; considering the telecom underlay network
OpenDHT	Open source	Simple interfaces to use
JXTA	Open source	A platform that enables any connected device on a network to exchange messages independently of the underlying network topology
Open Chord	Open source	An implementation of the Chord DHT
i3	Open source	Offers a rendezvous-based communication abstraction based on Chord; provide services like multicast, anycast, mobility and identifier stack
Overlay Weaver	Open source	Providing a common API for higher-level services such as DHT and multicast; designed in a modular way
Service routing layer	Commercial	A programmable overlay avoiding needless cost and complexity; combined with traditional IP layer

MaidSafe [80] which is an open source full Kademlia implementation with NAT traversal, FreePastry [81] which is an open source implementation of Pastry, JDHT [82] which is a simple Java based Distributed Hash Table, Khashmir [83] which is a Distributed Hash Table based on Kademila and written in Python, Mace [84] which is a C++ language extension and source-to-source compiler for building distributed systems, Dijjer [85] which is free P2P software that dramatically reduces the bandwidth needed to host large files. Akamai[TM] [86, 87] which is a famous Since so many implementations and plenty of researches based on them are exists, it gives a great push to the applications of DHT.

Chapter 4
DHT Applications

Like Pythagorean theorem and Newton Law, excellent theories are always simple and graceful. DHT, which performs so graceful that only two basic operations: *get* data from DHT and *put* data into DHT, is wildly used in many aspects of applications. In this chapter several DHT applications will be discussed.

DHT forms an infrastructure that can be used to build more complex services, such as multicast, anycast, distributed file systems, search, storage, content delivery network, file sharing, and communication. Going beyond the theory and implementations of DHT introduced in the previous chapters, we will focus on the applications of DHT in these aspects.

In the following sections, we will elaborate each application scenario, by first highlighting the motivations and the challenges. Then, we will illustrate how the problems are solved by DHT, or why the DHT method is suitable to the applications. For some applications, such as Multicast, DNS, and communication, we provide concrete examples to illustrate application scenarios. At the end of this chapter, we will summarize the role of DHT in the various applications.

4.1 Multicast

Multicast is one kind of message delivery that single source node sends a message in a single transmission to a group of destination nodes simultaneously. It can be widely used in public content broadcasting, voice and video conference, collaborative environments, games and so on. Compared with IP multicast requiring high cost to the infrastructure, overlay multicast or application multicast solely built on end-users can be implemented upon any infrastructures. The overlay multicast protocols are summarized in [88, 89] , and DHT is one way to meet the requirement.

Multicast is a distributed Internet-scale application. DTH has a property that each node only maintains partial information but can perceive any information in the whole system. It allows nodes in a multicast group only to maintain a little information about the group. Scribe [90], mDHT [91], XScribe [92],

H. Zhang et al., *Distributed Hash Table; Theory, Platforms and Applications*,
SpringerBriefs in Computer Science, DOI 10.1007/978-1-4614-9008-1_4,
© The Author(s) 2013

SplitStream [93] and Bayeux [94] all are the DHT based solutions of multicast. Here we briefly describe Scribe which is the best known DHT overlay multicast protocol.

Scribe is a scalable application-level Pastry based multicast infrastructure, which provides best-effort to deliver multicast messages. Scribe offers four API to its applications.

- CERATE: *create(credentials, groupId)*
  This function creates a group whose ID is groupId. The credentials are used for access control, which are provided to authenticate the node.
- JOIN: *join(credentials, groupId, messageHandler)*
  This function helps a node join the group with groupId. All the multicast messages this group receives are handled by the massageHandler.
- LEAVE: *leave(credentials, groupId)*
  This function causes the local node to leave the group whose ID is groupID.
- MULTICAST: *multicast(credentials, groupId, message)*
  This function broadcasts the message to the group with groupId.

Scribe constructs a multicast tree for each multicast group. When a node wants to create multicast group, it asks Pastry to route a CREATE message with a unique groupId. The message is routed to a node whose ID is closest to the groupId, which we call *rendezvous point*. The *rendezvous point* is the root of the multicast tree. Other nodes in the tree are called *forwarders*, which may or may not be the number of the group. Each forwarder maintains a *children table* for the group. When a node wants to join the group, it sends a JOIN message with the group's groupId, which would be routed toward the rendezvous point of the group. Each node along the route executes Scribe's forward method by the following way. If the local node is a forwarder of the multicast tree of the group, it accepts the node as a child and adds it to the child table. If not, the local node creates a child table and adds the source node. Then it sends a JOIN message along the route from the joining node to the rendezvous point and becomes a forwarder. The original JOIN message is terminated. Figure 4.1 illustrates the growth of a multicast tree in Scribe. In the beginning, the nodes in the multicast tree are "1100," "1101," "1110" and "0111," where "1100" is the root of the tree and the rendezvous point of the relevant group. When the node "0100" wants to join the group, it sends a JOIN message which is received by "1001." "1001" is not a forwarder of the tree, so it creates a child table for the group and inserts "0100" in it. Then it sends JOIN message to the next one "1110," which is already in the tree. "1110" inserts "1001" into its child table. Now the multicast tree possesses six nodes.

Each nonleaf node sends a heartbeat message to its children. When a child dose not receive the heartbeat message for a long time, it considers the parent failed and sends the JOIN message to repair the multicast tree. Particularly the root, which is the most important node in the tree, has an access control list and identifies the group creator. If it fails, the multicast tree is destroyed. Scribe replicates the information about the group on the k closest nodes to the root. The multiple replicas guarantee the group information still in Scribe when the root fails. If a leaf node leaves, it dose not influence other nodes.

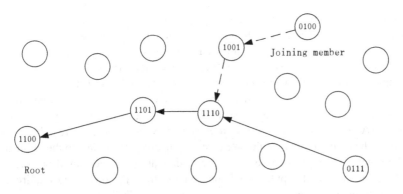

Fig. 4.1 The growth of a multicast tree in Scribe

When multicast sources disseminate messages to a group, it uses Pastry to locate the rendezvous point of the group. Then the root of the multicast tree of this group disseminates the messages along the tree. In this way, multicast messages are disseminated to the certain nodes. The randomization properties of Pastry ensure the balance of Scribe so that Scribe works well.

Multicast is the key technology of the streaming media application [95] where media data is constantly received by multiple nodes and is presented without downloading the whole media files. So the systems like SplitStream [93] construct multiple multicast trees on the DHT overlay to deliver the streaming packets.

4.2 Anycast

Anycast delivers the packets to the best node within a group according to different metrics such as latency, throughput, bandwidth and so on. Anycast can be used widely by many applications such as DNS and server automatic selection. The property of anycast that choosing the best target in a group automatically and transparently meets the requirement of CDN (Content Delivery Network) so well that it may be directly used in CDN. Like multicast, IP anycast is not supported well by the existed routers. Firstly, the upper protocols such as TCP cannot take advantage of it, because the packets from the same source node may be routed to the different destinations. Secondly, anycast is limited by the scalability that global anycast application may lead to huge and unmanageable routing tables. Thirdly, the IP anycast requires the support of routers.

Overlay anycast is the one way to solve the problem. Many overlay anycast architectures [96–99] employ proxies to manage the anycast group and record information about the hosts. So when an anycast request comes, the proxies choose the best target. Here many proxies existing in the network and much information should be stored and updated, so DHT may be used. Proxima [97] is a network coordinate based infrastructure providing lightweight and flexible anycast service. It deploys a DHT overlay which stores the synthetic coordinates of the hosts in Internet. When a client proposes an anycast request to a group of servers, it firstly gets the coordinates of servers from DHT. Then the RTT (Round-trip Time) can be calculated so that the nearest server could be selected. ChunkCast [100] is another solution. It focuses on the download from CDN where each object is partitioned into plenty of chunks. It builds a distributed directory tree for each object on top of Bamboo. All the chunks of the object represented by the bit vectors are published on the tree. The chunk lookup could be treated as an anycast message to the publishers of the object. Through the tree the nearest adequate publisher could be selected. Furthermore, if a better publisher hangs on the tree, the download requirement switches to the new peer.

Anycast is a key operation in many network scenarios, and IPv6 plans to support anycast directly [101]. However, it is not carried out yet so far, because of the non-uniform protocol standard. Meanwhile, both anycast of IP layer and anycast of overlay have their own strong points and defects, such as IP anycast is more efficient and overlay anycast is more flexible. Therefore, the DHT based anycast application is still worth exploiting in the era of IPv6.

4.3 DNS

The domain name system (DNS) is a naming system which translates domain names that are meaningful to humans into the IP addresses that are the numerical identifiers of computers in Internet. The unique IP identifiers are the foundation of the communications in Internet. Locating the host and selecting the route path all rely on the IP address. So DNS which provides the function of IP address lookup constitutes a critical component of Internet infrastructure. It is used extremely frequently everyday.

Currently DNS is a hierarchical system organized in a tree structure. A large distributed database stored the tuples of IP and domain name is maintained by the DNS servers all over the world. However, the tree structure has inherent limitations. The root of the tree is the bottleneck of the system. If it fails or the connections to the children are dropped, it would have great influence. Furthermore, the path of query is definite that the next hop is the father if the query is not matched locally, so DoS attack, which is the foremost problem with DNS, would be launched to disable the system easily. With its distributivity and safety needs, the DHT-style DNS is naturally proposed. The DHT-style DNS is a flat structure that all the nodes have the same weight and right. Every node has different next hop, so the DoS attack has much less influence. Moreover, DHT also provides efficient routing scheme which meets the requirement of quick response of DNS. Therefore, the DHT based DNS is another exciting application.

The cooperative domain name system (CoDoNS) [102] is a DHT based name service, providing high lookup performance, good load balance and fast propagation

Fig. 4.2 Proactive caching in beehive

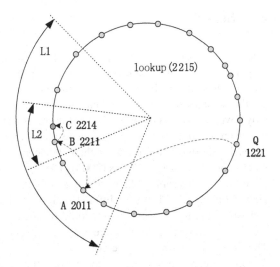

of updates. CoDoNS is built on Beehive [103], which is a proactive replication system that decreases the lookup hops to $O(1)$ in prefix-matching DHT like Pastry and Tapestry. Figure 4.2 illustrates the proactive caching of Beehive enhancing of prefix-digits like Pastry. Node Q queries an object with ID 2215. The first hop is A with ID 2011, then A transmits the query to $B(2211)$. At last B reroutes it to $C(2214)$ where the object 2215 is stored. There are three hops in classical Pastry. With proactive caching of Beehive, 2215 is cached among other nodes by the scheme that *level i* caching scheme ensures that the object can be located in i hops. Thus in Fig. 4.2, if $L2$ caching scheme is employed, all the nodes in $L2$ region cache replicas of the data. The path of the query would be decreased to 2 hops. If $L1$ is used, the query only need to be routed at 1 hop. The proactive replication reduces the hop number of query greatly which accelerates the response of DNS request. More outstanding ability of CoDoNS is defending the DoS attacks, that the excess query will be solved by increasing the number of replicas to spread the load. This is incompatible with the tree-style DNS architecture. Some works [104] had compared the two structures of DNS and evaluated the performance of the tree structure overwhelms the DHT except defending the attacks. But it only analyzes the basic DHT rather than some special DHT-style DNS such as CoDoNS.

DHT meets many requirements of DNS such as distribution, caching, replicas, and the put/get operations are suited to DNS. Moreover, nowadays the most serious threat to DNS system is attacks like DoS. So DHT would be a good choice to next generation name service of Internet.

4.4 Search

Search is one of the most important application nowadays. All of the searching corporations including Google[TM], Bing[TM], Yahoo![TM], Baidu[TM], and so on, have great influence to the Internet, even to the whole human society. Everyday hundreds of billions of search queries are launched. However these centralized search engines only cover a part of data of the whole Internet. Even Google[TM], the largest and best search engine in the world, only indexes about 1/10 of the total number of the Web pages. Furthermore, it does not include the deep Web pages, the number of which is much more than surface Web's (2008) [105].

In the recent years another type of search engine or technology enters our sight view. Compared with the traditional search engine, the distributed search engine consists of plenty of peers, which could be clients and servers at the same time. In the distributed search engine there is no center servers, where all the users are equal. The distributed search engine collects the power of the peers all over the world. The more peers are running in the network, the more powerful a distributed search engine is. Even if hundreds of millions of computers work as the distributed search engine peers, the engine would have more capability than the largest traditional search engine. From this point of view, the distributed search engine has infinite processing ability, unlimited storage space and very high bandwidth to solve all kinds of search queries form billions of users, which is much stronger than any centralized engine.

In the distributed searching, how to store and manage the huge distributed index database is the key challenge. DHT is one of the way to solve it. Every node maintains a part of the index database by the DHT allotment. The query and the index updating information would be routed to the exact node who maintains the related key. In this way the database and the queries are dispersed to all the nodes in the network. Furthermore, the DHT distributes the network load and query processing server, which could enhance the ability protecting from DoS attacks. Some distributed searching engines based on DHT are exploited. The following two engines are the most representative applications.

4.4.1 YaCy

YaCy[TM] [106] is a distributed open source search engine which is based on DHT. It is written by Java and supports Windows[TM], Linux and Mac OS[TM]. The latest version is YaCy[TM] 1.4 (May 2013). If a user wants to use YaCy[TM], firstly she installs a YaCy[TM] appliance. YaCy[TM] comprises web crawler, indexer, index library, user interface and a P2P network. The crawler harvests the web pages, which are the input of the indexer. The indexer indexes the pages and put them into the index library locally, which uses inverted index [107]. Then they will be merged into a global search index library which is an integrated NoSQL database shared for the YaCy[TM] network. When user queries an object, the result would be fetched from the global database.

YaCy^{TM} provides search API, which easily integrates YaCy^{TM} into other programming environments such as PHP or Perl. In the distributed search environment, YaCy^{TM} can provide superior privacy protection because the distributed architecture lead to no search logs and monitoring of search queries. The core concept of YaCy^{TM} is free, so YaCy^{TM} search means that information should be free access for everyone, and YaCy^{TM} itself is a GPL-licensed free engine. There are also several P2P search engines, such as Solandra [108] which is a real-time distributed search engine combined with Cassandra and Lucene [107], Elastic Search [109] which is built for the cloud and so on.

4.4.2 FAROO

FAROO^{TM} [110] is a commercial P2P search engine, which is the largest decentralized search engine that it has had more than 2.5 million peers as of May 2013.

FAROO^{TM} claims that it is the fastest P2P search engine with a mean response Time below one second, and it has more efficient index than any other P2P web search engines. Meanwhile, FAROO^{TM} supports multilingual search so that people all over the world could use it. FAROO^{TM} search architecture contains a distributed index and distributed crawlers, which are driven by users themselves. The capability of search engines depends on the power of crawlers. With the Internet growth, FAROO^{TM} search scale grows, and the distributed crawlers are more powerful. There are four application fields for FAROO^{TM} crawling. The first one is the attention based ranking. The ranking of the search results are voted by the visiting web pages, which is not so different from Page Rank. The second one is real-time search, from which users can find the results in the past month, past week, past day, even past hour that is too soon. The next one is indexing the deep web. Many web pages are crested on demand from databases which do not have incoming links. If crawlers want to crawl them, the engines must do more additional work. So these pages cannot be crawled by normal search crawlers. On the contrary, the capability of P2P search engines depends on the scale of network, which could be considered to be boundless. So FAROO^{TM} may do this work more easily than the normal search engines. The last one is personalization and behavioural targeted online advertising. This technology is based on click streams identified from network traffic, but got some buzz in the test by Phorm^{TM} [111].

FAROO^{TM} protects the privacy in two ways. One is by the architecture that it is hard to trace a user in the distributed environment. The other is encryption, i.e. all the queries are encrypted. FAROO^{TM} can be employed on multiple platforms such as PC, tablet, etc. The user interface of FAROO^{TM} is a web page like the normal search engine, and it is supported by most of the browsers, such as Internet Explorer^{TM}, Firefox^{TM}, Opera^{TM}, Safari^{TM}, Mozilla^{TM} and Chrome^{TM}. However it is not mentioned about Linux, maybe there will be a version of FAROO^{TM} for Linux users in the future.

Distributed search engine is a promising development. It provides more powerful processing ability, larger storage space, etc. It has higher scalability, more fault-tolerant and lower resource cost than traditional engine. Searching the whole Internet is a magnificent project, maybe uniting all the users on Internet is a more accordant way.

4.5   Storage

Nowadays more and more devices like cellular phones, televisions, even cars are capable of connecting to Internet. So storing personal data on the network to facilitate a variety of devices and to enhance the storage capability of them. An Internet-based storage system with strong persistence, high availability, scalability and security is required. Obviously the centralized methods is not a good way because it is lack of scalability and has the single point of failure problem. If the center fails, all the owners lose the capability to access their data which may cause inestimable losses. Besides, it is impossible to store all the data on one machine, though it is facility to management. Even in the cloud computing center which provides online storage functionality the data is distributed to tens of thousands of machines in a distributed way. Therefore, a distributed storage system seems to be a good choice to manage the gigantic storage.

How to organize so many kinds of data efficiently is the first hit. DHT with its wonderful structure is suitable to the distributed environment. DHT provides a high efficient data management scheme that each node in the system is responsible to a part of the data. It supports exact and quick routing algorithm to ensure users retrieving their data accurately and timely. Furthermore, replication and backup, fault-tolerant and data recovery, persistent access and update which are concerned in the storage area are not difficult to DHT. Recently many researches and systems [9, 112–115] are proposed to the development the DHT based storage. OceanStore [112] and PAST [113] are two famous solutions to the worldwide distributed storage.

4.5.1   OceanStore

OceanStore [112] is designed to provide global-scale persistent storage and continuous access to the information. Each object has a globally unique identifier (GUID), which is generated by the secure hash. OceanStore supports two types of data access control which are reader restriction and writer restriction. The former prevents unauthorized reads by encrypting all data in the system. The latter prevents unauthorized writes by signing all the write operations. For high availability and reliability, every object has multiple replicas dispersed in the network. However, the basic replicating scheme is sensitive because each object has a single root which is vulnerable to the single point of failure. OceanStore improves the scheme that it hashes each GUID with different salt values. Thus the results will be mapped to several root. Of course this modification will complicate the update of object. When one object is modified, all the replicas should be updated.

OceanStore also provides two routing mechanisms. First, a fast, probabilistic algorithm is used to find the target. The routing process is based on an attenuated Bloom filter that each node maintains a routing table consisting of Bloom filter vectors. The ith vectors merges all the Bloom filters of n-hop nodes. If it fails, a slower, deterministic algorithm will start.

DHT plays an important role in OceanStore, where all the GUID are mapped into the system in DHT manner. Each node maintains only a part of data and the roots of data are established by the Tapestry algorithm. Meanwhile, the deterministic algorithm also uses Tapestry, which enables the routing to be highly efficient. With Tapestry the target will be found within $O(\log n)$ hops, where n is the number of the nodes in the system.

4.5.2 PAST

PAST [113, 116] is an Internet-based, P2P global storage system, which is based on Pastry. It aims to provide strong persistence, high scalability, availability and security. Although Pastry already provides the functionality of data storage, PAST focuses on storage with much more consideration.

In PAST the length of node ID (nodeId) is 128 bits, and the length of file ID (fileId) is 160 bits. PAST exports three operations to its clients:

- *fileId* = Insert(*name, owner-credentials, k, file*)

  This function allows clients to store a file on k nodes in the network. The operation produces a fileId which is computed by the file name, the owner's public key and a randomly chosen salt.
- *file* = Lookup(*fileId*)

  This function helps node to retrieve a copy of the file with fileId if it exists in PAST.
- Reclaim(*fileId, owner-credentials*)

  This function reclaims the storage with fileId in PAST. Once this function is invoked, PAST no longer guarantees the file with fileId.

Smartcards are used to enhance the functionality of data storage in PAST. Quota management and security are two main issues. Each node in PAST has a limit of storage space. When node A receives an insert request and there is no enough space to store the file, file diversion is performed to achieve more global load balancing and storage capability of PAST. Node A chooses a node B in its leaf set which is not among the k closest nodes. Here "leaf set" is also called namespace set in Pastry, which are numerically closest to node A. A asks B to store the file and itself maintains a pointer to node B. In this way, A has diverted a copy to B.

Security is another important design consideration in PAST. PAST uses smartcard's public key to generate the node IDs, store files, and manage access to files. When a client tends to insert a file into PAST system, it first generates a file certificate and then routes the certificate with the file to the destinations. The file certificate allows each storing node to verify the authorization of the client and the file. When the client wants to reclaim the file, it generates a reclaim certificate with smartcard and sends it to the file storing nodes. With the reclaim certificate, the authentication of client can be verified.

Recently cloud storage is a hot topic, which studies online storage models where the data is originally stored on storage vendors rather than on host locally. Usually it is in the form of large-data seemingly centralized centers. In fact, all the data are stored in a distributed fashion on thousands of servers of data centers, though many of them are located in a same location. Some of the cloud storage solutions such as Cassandra [9, 117] choose DHT as the data distribution scheme. Hence, DHT is a promising technique that can be utilized in the cloud computing in many aspects.

4.6 CDN

Content delivery network (CDN) is a system that contains copies of data on various nodes. It improves the performance of access by caching the original servers. On the one hand, the caching scheme disperses the load to multiple replicate servers, which increases the access bandwidth. On the other hand, the request of user is routed to the nearest caching server, which reduces the latency of access. The traditional CDN copies the data from the original server to the multiple edge CDN servers. Nowadays there is a huge amount of data in Internet. It takes too much storage space to generate several mirrors of original servers. P2P CDNs based on DHT are proposed to lower the cost and higher the efficiency. In DHT based CDN, the replicas are organized by DHT technology. The data can be replicated dynamically. Figure 4.3 illustrates the deployment of DHT based CDN. A DHT network lies between server and clients. For example, when the data is requested, it is checked whether the data exists in the CDN by the DHT algorithm. If not, the relative CDN server firstly replicates the data from the original server. According to the number of requests, the replicas propagate reversely along the routing path with a timer. Once the data is requested again, the query stops at the first CDN server containing the relevant data and the timer on the server is reset. Thus the latency is reduced by shortening the path. If the timer expires, the data will be deleted from the caching server, which saves storage space for other more desired data. The more people require the data, the more replicas of it exist in the network. In this way the hot data has many copies in the network, and the response time is short, while the unpopular data has little copies in order to save the storage space and update bandwidth.

DHT based CDN optimizes the tradeoff of performance and cost. It provides the similar performance to the traditional CDN but costs much lower. Many DHT based CDNs have been exploited such as Bamboo [11], CodeeN [14],

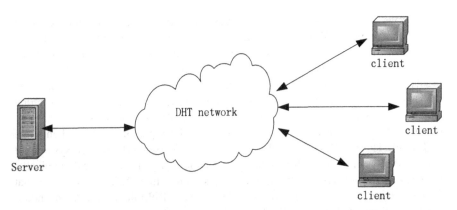

Fig. 4.3 The architecture of DHT based CDN

CoralCDN [7]. Because CDN is a kind of application that always is combined with other applications and for each of them is a platform of DHT implementation. Thus the detail of the introduction is shown in Sect. 3.1.

4.7 File Sharing

File sharing is a service providing access to the files and data stored in Internet. Here we discuss the global-scale file sharing application that files can be shared with the users all over the world. Users can find and download files in any corner of the world. By file sharing service of Internet, a lot of digital information such as multimedia documents, electronic books, software can be spread quickly and efficiently.

The first type of file sharing model is the traditional server system, where all the files are stored at servers, and users connect directly to the servers to fetch data. The most famous application is FTP service, which is still popular and used widely in nowadays. This completely centralized method that all the data aggregating at few nodes in the network has its own advantages. Firstly, the servers of files play too important roles in the file sharing. If the server fails, all the files on it would not be fetched. Besides, too much load and responsibility fall on servers' shoulders. Every server must maintain a large number of files, and tens of thousands of clients require files from one server at the same time. Moreover, it lacks of scalability that DoS attack is fatal to the servers. Therefore, dispersing the burden of servers is the key technologies of the file sharing. Then contents centralized directory P2P network comes out, the delegate of which is Napster [118], where the server stores the file indices rather than files themselves. The clients find out the file owners by the indices, and get the content from other peers. This technology that transmits the download burden from servers to peers takes significant progress, but the central index server still is the core and vulnerable point of the file sharing system. So Bittorrent introduces tracker servers, each of which maintains the lists of peers to distribute the key information.

However, this design arises some issues. For example, if you cannot connect to any tracker, you will be an isolated island in the Bittorrent network that hardly fetches any data. Fortunately, DHT completely overcomes the following three disadvantages of the server model. First of all, there is no server node. All the nodes in DHT network are equal. The failing of one or several nodes does not influence file sharing. When one node fails, other nodes will take change of the information it maintains. Secondly, the load is distributed to all the nodes. No one has too much load and responsibility. The network will be repaired quickly when some nodes fail. Thirdly, DHT is good at working in large-scale distributed systems. It endows the system with high scalability. The nodes can join and leave dynamically. So DHT is introduced into the file sharing systems, like DHT of Bittorrent that provides connection when the trackers fail, KAD of eMule where all the nodes are both clients and servers. DHT completely distributes the key information of the file sharing system. In DHT the server is not needed anymore. Each of the peers in the network is responsible for a part of file information. In the file sharing systems, Kademlia based DHT is the most popular DHT structure. Overnet [34], Bittorrent [36, 37], eDonkey, eMule [35] and the inheritances of them all develop their DHT networks based on Kademlia.

Another famous file sharing network is Freenet. Freenet at first is a distributed information storage system [119]. Now it is free software which lets users anonymously share files, browse an publish sites without censorship [120]. Compared with mainstream DHT file sharing networks like Bittorrent and eMule, Freenet aspires to share files without any auditing or tracking. Freenet is one of the earliest implements of DHT, the white paper of which was written in 1999 [121]. Until now the Freenet software has been downloaded more the two million times.

There are five design goals of Freenet, which are anonymity for both producers and consumers of information, deniability for stores of information, resistance to third parties' attempts to deny address to information, efficient dynamic storage/routing of information and decentralization of network function [122]. The last two aims could be solved by the DHT features, which also are the properties of other distributed softwares and platforms. Whereas the first three aims are extremely concerned about by Freenet. In Freenet, the privacy protection exists in every aspects of Freenet.

Freenet uses an unstructured architecture, files and nodes in Freenet have globally unique identifiers (GUIDs). In Freenet, there are two kinds of GUID keys. One is content-hash keys (CHK), which is generated by hashing the contents of the file stored in the network. Another is signed-subspace key (SSK), which sets up a personal namespace where anyone can read but only the owner can write. The SSK of a file is generated by the personal public-private key pair. Firstly user should summarize the file in a short description. Then she hashes the public key and the descriptive string independently. Thirdly, the two results are concatenated and hashed again. Lastly, the private key is used to sign the file providing an integrity check. In this way, if a user wants to retrieve a file from a subspace, she only needs the public key of the subspace and the description, from which the SSK could be recreated. However, if the user wants to update the file, the private key is required to generate a valid signature.

In the routing procedure every node forwards the message to the node closest to the target, and the routing is trained by the request processing. Freenet also tries its best to protect the privacy in this procedure. Massages are propagated through node-to-node chains. Every node on the chain only knows the predecessor and the successor, even dose not know the message's originator and the endpoint of the chain. Each link between two nodes is individually encrypted, so the routers do not know any other information about the packets except the neighbors. Furthermore, Freenet adds some mix-net route before normal routing to enhance security so that the path of the messages are monitored more hard. However, the high-level security brings some impact to the efficient of routing. Some improvements were proposed to solve this problem, maybe the small world theory is the most attractive way and several researches focus on it [123, 124].

More recently "darknet" [125, 126] as an important development of Freenet, enhances the security and privity of users for Freenet. It allows users only to connect to their friends, thus greatly reduce the vulnerability. At the mean time, users can still connect to any other peers of the network through their friends' connections, which makes it very difficult to block Freenet, which constructs a real "free world."

4.8 Communication

The communication system is a network that can connect remote people, such as telephone system, postal system. Here we focus on the communication based on Internet, which is mainly composed of Instant Messaging (IM) and Voice over IP (VoIP). IM connects two or more people by real-time text-based communication. The classical delegates are MSN^{TM}, QQ^{TM}. VoIP is the same as IM, except that the form of communicating information is voice rather than text. In this area Skype^{TM} is the leader. Nowadays the two services are often intertwined in the same application software, which users can chat through MSN^{TM} and QQ^{TM}, and also can send text information by Skype^{TM}.

Recently the majority of IM and VoIP systems are centralized, a user agent installed on the client's machine connects to a central server. All the information such as buddy list, state of online or outline, is stored on the server, even the communication messages between two clients are transmitted by the central server. Like all the centralized application, the central server is the bottleneck and weak point. Furthermore, some scenarios, such as prohibiting contacting an outside server and ephemeral groups, also requires no central server communication system.

In the distributed communication system, DHT plays an important role that has been used in the commercial system. Skype^{TM}, which is one of the most famous applications in VoIP and IM areas, employs DHT in its network. In the following we will introduce the distributed architecture of Skype^{TM}.

Skype^{TM} is composed of the ordinary hosts, the super nodes and the login server [127]. The login server is the only centralized component in Skype^{TM} network. It is in charge of the usernames and passwords about Skype^{TM} users.

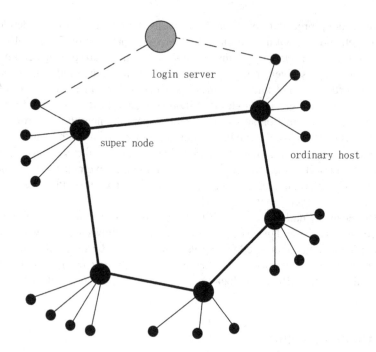

Fig. 4.4 The architecture of Skype™ network. It is composed of the login server, the ordinary hosts and the super nodes

All the users must connect to the login server for authentication in the login phase. The ordinary hosts are the machines or devices that can run Skype™ application. Tens of millions of hosts are widely distributed in the world. They possess different capabilities of CPU, memory, storage and bandwidth. Some of the hosts with public IP addresses and strong capability are chosen as the super nodes. Each super node manages a group of ordinary hosts. The super nodes compose the backbone of Skype™ network. Almost all the information is transmitted in the super node network, except the first hop (from source host to the super node in charge of the source) and the last hop (from super node to the destination). Figure 4.4 shows these three entities in the Skype™ network. The architecture of Skype™ is a hierarchical network, where a group of ordinary hosts and a super node construct a small-scale centralized network, and the super nodes make up a distributed network.

Except the login information, all of other data is stored distributedly in the super node network. With its private Global Index (GI) technology, the hosts can search for users accurately and efficiently. We are not sure about the relationship of GI and DHT, but obversely DHT can be used to store and retrieve data in the distributed environment. In SOSIMPLE [128] DHT has been used for lookups. Meanwhile, since each super node maintains a group of ordinary hosts, in DHT the division of host groups can be determined by the IDs. In other words, the group can be shaped by the relationship of nodes' IDs in DHT, e.g. the precursor-successor relationship is formed according to the distance between the super node and the host. DHT provides high scalability and can work with churning, so it is easy to solve the problem of the group member's change.

DHT also can be used to interconnecting the different IM networks. CrossTalk [129] constructs a public gateway DHT network to store the presence and name to gateway mappings. Each IM network has gateways which are responsible to the information format conversion. Users will announce their status in the DHT and get the foreign buddies from it. When two users in different IM systems talk to each other, the sponsor will locates the gateway of the recipient in DHT. Hence, DHT fulfills itself as a distributed storage and data location network.

IM and VoIP in essential are communication among end points. In this aspect it does not need the participation of the servers. Therefore, besides using DHT's storage capability, the communication may also be the DHT style without the central server, which is more direct and suitable to meet the end-to-end requirement.

4.9 Conclusion

As discussed above, DHT can be applied in many aspects to improve the performance of communication networks. Table 4.1 summarizes the applications DHT has been developed in. Multicast usually needs to build a multicast tree structure to propagate information. Compared with the regular Internet routing algorithms, which build the routing table by the local information, routers of DHT know the architecture naturally, without broadcasting to perceiving the network as Distance Vector algorithm and Link State algorithm. When the destination address is determined, every router knows the next hop. Each hop makes the message closer to the target. Therefore, if the root of the multicast tree is determined, the tree could be constructed based on the DHT routing scheme. DHT also can be used in anycast applications. Some projects [97] employ the scalability of storage and accuracy of data location of DHT to store the information about the servers, some [100] construct the directory tree for each object. The DHT routing algorithms ensure that the next hop is closer to the target so that all the path will be aggregated at the root of the tree. The data is stored along the routing path. When a query is transmitted to the node of the tree, the nearest adequate publisher could be selected. For DNS, DHT provides another flat structure solution, where all the DNS servers are equal. So it can conquer the root bottleneck of the classical tree structure. In the searching area, DHT could be used to build the distributed search engine. The index database is distributively stored on the nodes in the network. Each node maintains a part of the database allotted by DHT. Meanwhile, the search queries are also dispersed to the whole network. DHT combines all the nodes in the network to construct a powerful search engine, which could be stronger and stronger with the increment of the users. In storage area, the distributed network storage system could develop the allotment strategy by DHT. With the DHT's help the distributed storage system could be organized by itself. Every node knows the data it should maintain

Table 4.1 Comparison of DHT applications

Application	The mainly used properties of DHT	Advantages by introducing DHT
Multicast	The route table is determined by DHT algorithms, not by broadcasting. Each hop makes the message closer to the target	Easy to build a multicast tree
Anycast	Distributed storage and efficient accurate data location; Each hop makes the message closer to the target	High storage scalability and rapid position; Easy to build a directory tree
DNS	The equality of DHT nodes	Conquer the bottleneck of root of tree structure and the vulnerability for DoS
Search	Distributed storage and efficient accurate data location	Enhance the scalability and ability of engine
Storage	Distributed storage and efficient accurate data location	High storage scalability; Prevent the single point of failure
CDN	Distributed storage and efficient accurate data location	Save the storage space
File sharing	Distributed storage and efficient accurate data location	High storage scalability; Prevent the single point of failure
Communication	Distributed storage; The determined network structure	Prevent the single point of failure; Improve the performance of long distance communication

naturally, without any announcement of the third party. For file sharing, DHT plays the similar role as what in the storage. DHT allocates each node a part of data to maintain. The routing path could be ensured only by knowing the target's ID. For communication, DHT provides the interfaces that most of the information of the users is stored in the distributed way, and provides a distributed way to construct the network architecture, which has shown the advantages especially in cross-border long distance communication.

From the analysis above, DHT applications have two properties. One is distributivity. DHT owns an allotting scheme that distributes the data to the nodes in a distributed environment. All the nodes in the system can work together harmoniously. Without the assignment of the central server, every node knows which part of data it should maintain. The location of data and the routing path are determined only by the ID. Because of the clarity of the DHT network structure, the routing table is predictable without broadcasting. This ability of well organizing resources and nodes makes DHT suitable for the distributed applications. The other is large scale. DHT provides high scalable storage and efficient routing algorithms, which are sufficient to meet the request of the global-scale applications. In a distributed system with tens of thousands of nodes, the failing of nodes is normal state. DHT has efficient ways to solve this problem. In the churning environment DHT performs well.

Except the applications above, there are also many other aspects DHT can play an important role, such as network games [130], device-tracking [131, 132], etc. Resource sharing is a basic service of DHT. Nowadays the data sharing system based on DHT has been applied widely. The application of data sharing is so mature that music, movies, books and many kinds of multimedia resources are shared globally. Currently another important resource called computing resource becomes popular. Computing resource also is a valuable resource [133] that has a huge commercial market. Cloud computing is the most striking one that sells computing resources to renters. Cloud computing in this phase shows in the form of centralization, which limits the scale of it. Distributed computing and Peer-to-Peer computing have appeared for a long time, some projects such as SETI@home [134] have shown the unparalleled power of the distributed computing. However, SETI@home also needs a central server to segment the task which is a very heavy work. Clients do not communicate to each other. If someday DHT is used in the computing, and the computing develops to a pure distributed style, the computing power may be enhanced to a novel stage.

Chapter 5
Summary

In the above chapters, we summarize the theory, platforms and applications of DHT. In theory, there are plenty of variants of DHT, which provide many choices for constructing DHT-based platforms and applications. Meanwhile, a number of platforms of DHT have been built for several years, both in academic area and commercial area. These platforms provide an efficient way to construct applications. The direct reason of prosperity of theory and platforms is the wide range of uses of DHT. On the one hand, from the summary in the above chapters, DHT is widely used in many aspects, which motivates the research for DHT. On the other hand, there are lots of studies contributed on DHT, which is convenient for developing the applications. All in all, DHT is a promising technology and has much influence on the development of networks.

However, DHT also has its limitations. First of all, DHT is not good at calculating global statistics. In C/S model, the server knows the global states of the system, so it is easy to get global information about the system. For example, if a user wants to know the number of users, it is easy to get the information from server. On the contrary, in DHT-based systems there is no central node. Each node only knows a portion of information. When a global statistic is required, all the nodes (or multiple nodes) in the system have to join in the process of calculation, which has additional communication cost and delay. Meanwhile, during the process, the state of the node in the system may change over time, e.g., new node's joining, node's leaving or failing, which makes the result hard to represent the real-time status of system accurately. Thus in this system, the solution for commuting global statistics navigates a tradeoff among accuracy, cost, performance and robustness [135]. Normally three kinds of methods are used in aggregation [136], namely, gossip-based method [137], tree-based method [138] and mixture method [139]. However, due to the lack of central node in DHT-based distribution system, it is not as convenient as centralized model to calculate the global aggregate statistics.

Moreover, complex searches are not suitable for DHT-based applications. In search area, for exact match, there is no problem for DHT. However, complex searches, including range search and substring search, are invalid in DHT. Due to the

H. Zhang et al., *Distributed Hash Table; Theory, Platforms and Applications*,
SpringerBriefs in Computer Science, DOI 10.1007/978-1-4614-9008-1_5,
© The Author(s) 2013

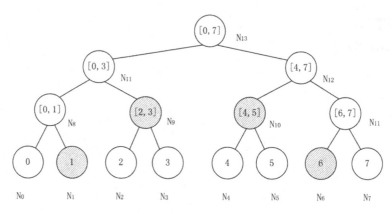

Fig. 5.1 The range search tree in CANDy

hash mapping, DHT changes the properties of the original resources. The resources with little difference are mapped to quite different values. Plenty of research work has been contributed in this area. For example, in [140] substring searches are realized by the exact-match facility of hash. The original string and query string are split into several n-length substrings, so the exact string match can be used directly by matching the n-length substrings. Here the typical n is 2 or 3. For example, the filename of a file with ID I is "Mona Lisa," which could be split into seven trigrams: Mon, ona, na%, a%L, %Li, Lis, isa ("%" represents the space character). For each 3-gram g_i, the pair (g_i, I) is inserted into the hash index, keyed by g_i. When a substring query is required (e.g. "Mona Li"), it also be split into 3-length substrings to search. This scheme indeed solves the problem of substring search, but it costs too much. For the range search, "range search tree" is a popular method [141], which uses the tree structure to organize the data. Figure 5.1 is an example about range search in CANDy [141]. The value space $V = \{0, 1, 2, \cdots, 7\}$, which is stored in a 4-level tree. The leaves of the tree comprise the individual values. The parent contains the values of its children. In this case, the first leaf N_1 from the left has the value 0, and its parent N_8 covers the subrange from 0 to 1. The root of the tree covers the full value space V. When a range request $\{1, 2, \cdots, 6\}$ comes, four subqueries are generated: the first for 1 from the node N_1, the second for $[2, 3]$ from N_9, the third for $[4, 5]$ from N_{10} and the fourth for 6 from N_6. Although there are plenty of researches in this area [142–144], all of them incur high storage cost or computational complexity. The property of hash function makes that the complex search is a difficult issue.

Last but not the least, security is another concerned issue in DHT-based research. In C/S model, the server knows the global information of the system, and is the most important node, which is more trustable than the clients. In DHT environment each node only knows a small subset of other nodes, and all the nodes are traded the same. This property makes it difficult to prevent malicious nodes from joining the system. The most studied DHT attacks in the literature are Sybil attack, Eclipse attack and routing/storage attacks [145]. All of them are based on lack of authentification. A node is not able to judge others to be honest or not, which makes the security scheme in DHT-based applications is much more difficult than C/S based applications, or other applications that have nodes playing the dominant role.

References

1. W. Litwin, Marie-Anna Neimat, and D. A. Schneider. Lh*-a scalable, distributed data structure. In *ACM Transactions on Database Systems*, volume 21, December 1996.
2. Distributed hash table. http://en.wikipedia.org/wiki/Distributed_hash_table.
3. Eng Keong Lua, Jon Crowcroft, Marcelo Pias, Ravi Sharma, and Steven Lim. A survey and comparison of peer-to-peer overlay network schemes. *IEEE Communications Surveys and Tutorials*, 7:72–93, 2005.
4. Stephanos, Androutsellis-Theotokis, and Diomidis Spinellis. A survey of peer-to-peer content distribution technologies. *ACM Computing Surveys*, 36(4):335–371, December 2004.
5. D. Karger, E. Lehman, T. Leighton, R. Panigrahy, M. Levine, and D. Lewin. Consistent hashing and random trees: distributed caching protocols for relieving hot spots on the world wide web. In *STOC '97 Proceedings of the twenty-ninth annual ACM symposium on Theory of computing*, pages 654–663, May 1997.
6. Ion Stoica, Robert Morris, David Karger, M. Frans Kaashoek, and Hari Balakrishnan. Chord: A scalable peer-to-peer lookup service for internet applications. In *SIGCOMM '01 Proceedings of the 2001 conference on Applications, technologies, architectures, and protocols for computer communications*, pages 149–160, August 2001.
7. Coralcdn. http://www.coralcdn.org/.
8. Hazelcast. http://www.hazelcast.com.
9. Avinash Lakshman and Prashant Malik. Cassandra: a decentralized structured storage system. *ACM SIGOPS Operating Systems Review*, 44:35–40, April 2010.
10. Websphere extreme scale. http://www-01.ibm.com/software/webservers/appserv/extremescale.
11. Bamboo. http://bamboo-dht.org.
12. Giuseppe DeCandia, Deniz Hastorun, Madan Jampani, Gunavardhan Kakulapati, Avinash Lakshman, Alex Pilchin, Swaminathan Sivasubramanian, Peter Vosshall, and Werner Vogels. Dynamo: Amazon's highly available key-value store. In *Proceedings of twenty-first ACM SIGOPS symposium on Operating systems principles*, 2007.
13. Guangyu Shi, Jian Chen, Hao Gong, Lingyuan Fan, Haiqiang Xue, Qingming Lu, and Liang Liang. A dht key-value storage system with carrier grade performance. In *Euro-Par 2009 Parallel Processing*, volume 5704, pages 361–374, 2009.
14. Codeen. http://codeen.cs.princeton.edu/.
15. Gnunet: Gnu' framework for secure peer-to-peer networking. https://gnunet.org/.
16. Filipe Araujo and Luis Rodrigues. Survey on distributed hash tables, 2006. https://estudogeral.sib.uc.pt/handle/10316/13603.
17. Stephanos Androutsellis-Theotokis and Diomidis Spinellis. A survey of peer-to-peer content distribution technologies. *ACM Computing Surveys*, 36:335–371, December 2004.

H. Zhang et al., *Distributed Hash Table; Theory, Platforms and Applications*,
SpringerBriefs in Computer Science, DOI 10.1007/978-1-4614-9008-1,
© The Author(s) 2013

18. Siamak Sarmady. A survey on peer-to-peer and dht, 2010. http://www.sarmady.com/siamak/papers/p2p-290808-2.pdf.
19. Zhang YiMing, Lu XiCheng, and Li DongSheng. Survey of dht topology construction techniques in virtual computing environments. *Science China Information sciences*, 54(11):2221–2235, November 2011.
20. Sean Rhea, Dennis Geels, Timothy Roscoe, and John Kubiatowicz. Handling churn in a dht. In *Proceedings of the USENIX Annual Technical Conference*, June 2004.
21. Us secure hash algorithm 1 (sha1), 2001. http://http://www.ietf.org/rfc/rfc3174.txt.
22. Paola Flocchini, Amiya Nayak, and Ming Xie. Enhancing peer-to-peer systems through redundancy. *IEEE Journal on Selected Areas in Communications*, 25(1):15–24, January 2007.
23. Yuh-Jzer Joung and Jiaw-Chang Wang. $chord^2$:a two-layer chord for reducing maintenance overhead via heterogeneity. *Computer Networks*, 51(3):712–731, February 2007.
24. M. Frans Kaashoek and David R. Karger. Koorde: A simple degree-optimal distributed hash table. *Lecture Notes in Computer Science*, 2735:98–107, 2003.
25. De bruijn graph. http://en.wikipedia.org/wiki/De_Bruijn_graph.
26. G. Cordasco, L. Gargano, A. Negro, V. Scarano, and M. Hammar. F-chord: Improved uniform routing on chord. *Networks*, 52(4):325–332, June 2008.
27. Moni Naor and Udi Wiede. Know thy neighbor's neighbor: Better routing for skip-graphs and small worlds. In *Peer-to-Peer Systems III*, volume 3279, pages 269–277, 2005.
28. G. S. Manku. The power od lookahead in small-world routing networks. In *STOC*, 2004.
29. Gurmeet Singh Manku, Moni Naor, and Udi Wieder. Know thy neighbor's neighbor: The power of lookahead in randomized p2p networks. In *Proceedings of the thirty-sixth annual ACM symposium on Theory of computing*, pages 54–63, 2004.
30. Prasanna Ganesan and Gurmeet Singh Manku. Optimal routing in chord. In *Proceedings of the fifteenth annual ACM-SIAM symposium on Discrete algorithms*, pages 176–185, 2004.
31. Sylvia Ratnasamy, Paul Francis, Mark Handley, Richard Karp, and Scott Schenker. A scalable content-addressable network. In *SIGCOMM '01 Proceedings of the 2001 conference on Applications, technologies, architectures, and protocols for computer communications*, pages 161–172, August 2001.
32. Daishi Kato. Gisp: Global information sharing protocol—a distributed index for peer-to-peer systems. In *Second International Conference on Peer-to-Peer Computing*, pages 65–72, September 2002.
33. Petar Maymounkov and David Mazières. Gisp: Global information sharing protocol—a distributed index for peer-to-peer systems. In *1st International Workshop on Peer-to-Peer Systems*, March 2002.
34. Overnet. http://www.overnet.org/.
35. emule. http://www.amule.org/.
36. emule. http://www.bittorrent.com.
37. Johan Pouwelse, Pawel Garbacki, Dick Epema, and Henk Sips. The bittorrent p2p file-sharing system: Measurements and analysis. *Lecture Notes in Computer Science*, 3640:205–216, 2005.
38. D. Stutzbach and R. Rejaie. Improving lookup performance over a widely-deployed dht. In *25th IEEE International Conference on Computer Communications*, 2006.
39. Di Wu, Ye Tian, and Kam Wing Ng. Analytical study on improving lookup performance of distributed hash table systems under churn. In *Sixth IEEE International Conference on Peer-to-Peer Computing*, September 2006.
40. Zhonghong Ou, Erkki Harjula, Otso Kassinen, and Mika Ylianttila. Performance evaluation of a kademlia-based communication-oriented p2p system under churn. *Computer Networks*, 54(5):689–705, April 2010.
41. Hun J. Kang, Eric Chan-Tin, Nicholas J. Hopper, and Yongdae Kim. Why kad lookup fails. In *IEEE P2P'09*, September 2009.
42. Andreas Binzenhofer and Holger Schnabel. Improving the performance and robustness of kademlia-based overlay networks. *Kommunikation in Verteilten Systemen (KiVS)*, pages 15–26, 2007.

43. Antony Rowstron and Peter Druschel. Pastry: Scalabel, distributed object location and routing for large-scale peer-to-peer systems. In *IFIP/ACM International Conference on Distributed Systems Platforms*, pages 329–350, November 2001.

44. Ben Y. Zhao, Ling Huang, Jeremy Stribling, Sean C. Rhea, Anthony D. Joseph, and John D. Kubiatowicz. Tapestry: A resilient global-scale overlay for service deployment. In *IEEE Journal on Selected Areas in Communications*, volume 22, 2004.

45. An architecture for ip address allocation with cidr, September 1993. http://merlot.tools.ietf.org/html/rfc1518.

46. Dahlia Malkhi, Moni Naor, and David Ratajczak. Viceroy: A scalabel and dynamic emulation of the butterfly. In *Proceedings of the twenty-first annual symposium on Principles of distributed computing*, pages 183–192, 2002.

47. Elena Meshkovaa, Janne Riihijarvia, Marina Petrovaa, and Petri Mahonen. A survey on resource discovery mechanisms, peer-to-peer and service discovery frameworks. *Computer Networks*, 52(11):2097–2128, August 2008.

48. C. G. Plaxton, R. Rajaraman, and A. W. Richa. Accessing nearby copies of replicated objects in a distributed environment. *Theory of Computing Systems*, 32(3):241–280, 1999.

49. Ryan Huebsch, Joseph M. Hellerstein, Nick Lanham, Boon Thau Loo, Scott Shenker, and Ion Stoica. Querying the internet with pier. In *Proceedings of the 29th international conference on Very large data bases*, volume 29, pages 321–332, 2003.

50. Opendht. http://opendht.org/.

51. Planetlab. http://www.planet-lab.org/.

52. Michael J. Freedman, Eric Freudenthal, and David Mazieres. Democratizing content publication with coral. In *Proceedings of the 1st conference on Symposium on Networked Systems Design and Implementation*, 2004.

53. Michael J. Freedman and Princeton University. Experiences with coralcdn: A five-year operational view. In *Proceedings of the 7th USENIX conference on Networked systems design and implementation*, 2010.

54. Sean Rhea, Brighten Godfrey, Brad Karp, John Kubiatowicz, Sylvia Ratnasamy, Scott Shenker, Ion Stoica, and Harlan Yu. Opendht: A public dht service and its uses. In *Proceedings of the 2005 conference on Applications, technologies, architectures, and protocols for computer communications*, 2005.

55. Jxta: The language and platform independent protocol for p2p networking. http://jxta.kenai.com.

56. D. Brookshier, D. Govoni, N. Krishnan, and J. C. Soto. *JXTA: Java P2P Programming*. Daniel Brookshier Darren Govoni Navaneeth Krishnan Juan Carlos Sotoen, 2002.

57. Jxta v2.0 protocols specification. http://jxta.kenai.com/Specifications/JXTAProtocols2_0.pdf.

58. Bernard Traversat, Mohamed Abdelaziz, and Eric Pouyoul. Project jxta: A loosely-consistent dht rendezvous walker, 2004. http://citeseerx.ist.psu.edu/viewdoc/download?doi=10.1.1.3.9419&rep=rep1&type=pdf.

59. Gnunet. http://en.wikipedia.org/wiki/GNUnet.

60. Christian Grothoff. The gnunet dht. http://grothoff.org/christian/teaching/2011/2194/dht.pdf.

61. Krista Bennett, Christian Grothoff, Tzvetan Horozov, Ioana Patrascu, and Tiberiu Stef. Gnunet - a truly anonymous networking infrastructure. In *Proc. Privacy Enhancing Technologies Workshop*, 2002.

62. Openchord. http://www.uni-bamberg.de/en/fakultaeten/wiai/faecher/informatik/lspi/bereich/research/software_projects/openchord/.

63. S. Kaffille and K. Loesing. *Open Chord version 1.0.4 User's Manual*, 2007. http://www.uni-bamberg.de/fileadmin/uni/fakultaeten/wiai_lehrstuehle/praktische_informatik/Dateien/Forschung/open-chord_1.0.4_manual.pdf.

64. Hazelcast documentation. http://www.hazelcast.com/documentation.jsp.

65. Patrick Hunt, Mahadev Konar, Flavio P. Junqueira, and Benjamin Reed. Zookeeper: wait-free coordination for internet-scale systems. In *Proceeding USENIXATC'10 Proceedings of the 2010 USENIX conference on USENIX annual technical conference*, 2010.

66. Internet indirection infrastructure. http://i3.cs.berkeley.edu/.

67. Ion Stoica, Daniel Adkins, Shelley Zhuang, Scott Shenker, and Sonesh Surana. Internet indirection infrastructure. In *SIGCOMM*, 2002.
68. Overlay weaver. http://overlayweaver.sourceforge.net/.
69. Kazuyuki Shudo, Yoshio Tanaka, and Satoshi Sekiguchi. Overlay weaver: An overlay construction toolkit. *Computer Communications*, 31(2):402–412, February 2008.
70. Apache software foundation. http://en.wikipedia.org/wiki/Apache_Software_Foundation.
71. Fay Chang, Jeffrey Dean, Sanjay Ghemawat, Wilson C. Hsieh, Deborah A. Wallach, Mike Burrows, Tushar Chandra, Andrew Fikes, and Robert E. Gruber. Bigtable: A distributed storage system for structured data. In *Proceedings of the 7th Conference on USENIX Symposium on Operating Systems Design and Implementation*, volume 7, pages 205–218, 2006.
72. Naohiro Hayashibara, Xavier Defago, Rami Yared, and Takuya Katayama. The φ accrual failure detector. In *Proceedings of the 23rd IEEE International Symposium on Reliable Distributed Systems*, 2004.
73. Cassandra. http://en.wikipedia.org/wiki/Apache_Cassandra.
74. Cassandra. http://cassandra.apache.org/.
75. IBM. *Scalable, Integrated Solutions for Elastic Caching Using IBM WebSphere eXtreme Scale*, 2011. Redbooks.
76. Dynamo: A flawed architecture. http://news.ycombinator.com/item?id=915212.
77. Bruce S. Davie and Jan Medved. A programmable overlay router for service provider innovation. In *Proceedings of the 2nd ACM SIGCOMM workshop on Programmable routers for extensible services of tomorrow*, 2009.
78. Sharkypy. http://www.heim-d.uni-sb.de/~heikowu/SharkyPy/.
79. Chimera. http://current.cs.ucsb.edu/projects/chimera/.
80. Maidsafe. http://www.maidsafe.net/maidsafe-dht.asp.
81. Freepastry. http://www.freepastry.org/FreePastry/.
82. Jdht. http://dks.sics.se/jdht/.
83. Khashmir. http://khashmir.sourceforge.net/.
84. Charles Killian, James W. Anderson, Ryan Braud, Ranjit Jhala, and Amin Vahdat. Mace: Language support for building distributed systems. In *Proceedings of the 2007 ACM SIGPLAN conference on Programming language design and implementation*, 2007.
85. Dijjer. http://www.dijjer.org.
86. Akamai. http://www.akamai.com/.
87. Ao-Jan Su, David R. Choffnes, Aleksandar Kuzmanovic, and Fabian E. Bustamante. Drafting behind akamai. In *SIGCOMM'06*, 2006.
88. D. M. Moen. Overview of overlay multicast protocols. http://bacon.gmu.edu/XOM/pdfs/Multicast%20Overview.pdf.
89. Stefan Birrer and Fabian E. Bustamante. A comparison of resilient overlay multicast approaches. *IEEE Journal on Selected Areas in Communications*, 25(9):1695–1705, 2007.
90. Miguel Castro, Peter Druschel, Anne Marie Kermarrec, and Antony Rowstron. Scribe: A large-scale and decentralized application-level multicast infrasstructure. In *IEEE Journal on Selected Areas in Communications*, volume 20, October 2002.
91. Jae Woo Lee, Henning Schulzrinne, Wolfgang Kellerer, and Zoran Despotovic. mdht: Multicast-augmented dht architecture for high availability and immunity to churn. In *Proceedings of the 6th IEEE Conference on Consumer Communications and Networking Conference*, 2009.
92. Andrea Passarella, Franca Delmastro, and Marco Conti. Xscribe: a stateless, cross-layer approach to p2p multicast in multi-hop ad hoc networks. In *Proceedings of the 1st international workshop on Decentralized resource sharing in mobile computing and networking*, 2006.
93. Miguel Castro, Peter Druschel, Anne-Marie Kermarrec, Animesh Nandi, Antony Rowstron, and Atul Singh. Splitstream: high-bandwidth multicast in cooperative environments. *ACM SIGOPS Operating Systems Review*, 37(5):298–313, December 2003.

94. Shelley Q. Zhuang, Ben Y. Zhao, Anthony D. Joseph, Randy H. Katz, and John D. Kubiatowicz. Bayeux: An architecture for scalable and fault-tolerant wide-area data dissemination. In *Proceedings of the 11th international workshop on Network and operating systems support for digital audio and video*, 2001.

95. W.-P. Ken Yiu, Xing Jin, and S.-H. Gary Chan. Challenges and approaches in large-scale p2p media streaming. *IEEE Multimedia*, 14(2):50–59, 2007.

96. Michael J. Freedman, Karthik Lakshminarayanan, and David Mazières. Oasis: Anycast for any service. In *Proceedings of the 3rd conference on Networked Systems Design and Implementation - Volume 3*, 2006.

97. Guodong Wang, Yang Chen, Lei Shi, Eng Keong Lua, Beixing Deng, and Xing Li. Proxima: towards lightweight and flexible anycast service. In *Proceedings of the 28th IEEE international conference on Computer Communications Workshops*, 2009.

98. Hitesh Ballani and Paul Francis. Towards a global ip anycast service. In *Proceedings of the 2005 conference on Applications, technologies, architectures, and protocols for computer communications*, 2005.

99. Hitesh Ballani and Paul Francis. Analysis of an anycast based overlay system for scalable service discovery and execution. *Computer Networks*, 54(1):97–111, January 2010.

100. Byung gon Chun, Peter Wu, Hakim Weatherspoon, and John Kubiatowicz. Chunkcast: An anycast service for large content distribution. In *In Proceedings of the 5th International Workshop on Peer-to-Peer Systems*, 2006.

101. Ip version 6 addressing architecture. http://tools.ietf.org/html/rfc4291.

102. Venugopalan Ramasubramanian and Emin Gün Sirer. The design and implementation of a next generation name service for the internet. In *Proceedings of the 2004 conference on Applications, technologies, architectures, and protocols for computer communications*, 2004.

103. Venugopalan Ramasubramanian and Emin Gün Sirer. Beehive: Exploiting power law query distributions for o(1) lookup performance in peer to peer overlays. symposium on networked systems design and implementation. In *Proceedings of the 1st conference on Symposium on Networked Systems Design and Implementation - Volume 1*, 2004.

104. V. Pappas, D. Massey, A. Terzis, , and L. Zhang. A comparative study of the dns design with dht-based alternatives. In *INFOCOM*, 2006.

105. Fang Qiming, Yang Guangwen, Wu Yongwei, and Zheng Weimin. P2p web search technology. *Journal of Software*, 19(10):2706–2719, October 2008.

106. Yacy. http://yacy.de/.

107. B. Goetz. The lucene search engine: Powerful, flexible, and free, 2002. http://www.javaworld.com/javaworld/jw-09-2000/jw-0915-lucene.html?page=1.

108. Solandra. https://github.com/tjake/Solandra.

109. Elastic. http://www.elasticsearch.org/.

110. Faroo. http://www.faroo.com.

111. Phorm. http://en.wikipedia.org/wiki/Phorm.

112. John Kubiatowicz, David Bindel, Yan Chen, Steven Czerwinski, Patrick Eaton, Dennis Geels, Ramakrishna Gummadi, Sean Rhea, Hakim Weatherspoon, Chris Wells, and Ben Zhao. Oceanstore: An architecture for global-scale persistent storage. *ACM SIGOPS Operating Systems Review*, 34(5):190–201, November 2000.

113. Peter Druschel and Antony Rowstron. Past: A large-scale, persistent peer-to-peer storage utility. In *Eighth Workshop on Hot Topics in Operating Systems*, pages 75–80, May 2001.

114. Marcel Karnstedt, Kai-Uwe Sattler, Martin Richtarsky, Jessica Muller, Manfred Hauswirth, Roman Schmidt, and Renault John. Unistore: Querying a dht-based universal storage. *IEEE 23rd International Conference on Data Engineering*, 1:1503–1504, 2007.

115. Martin Raack. Okeanos: Reconfigurable fault-tolerant transactional storage supporting object deletions. In *High Performance Computing (HiPC), 2010 International Conference on*, pages 1–10, december 2011.

116. Antony Rowstron and Peter Druschel. Storage management and caching in past, a large-scale, persistent peer-to-peer storage utility. In *SOSP '01 Proceedings of the eighteenth ACM symposium on Operating systems principles*, pages 188–201, 2001.

117. Qinlu He, Zhanhuai Li, and Xiao Zhang. Study on dht based open p2p cloud storage services systems. In *2010 International Conference on Computational and Information Sciences*, 2010.

118. Napster. http://en.wikipedia.org/wiki/Napster.

119. Ian Clarke, Scott G. Miller, Theodore W. Hong, Oskar Sandberg, and Brandon Wiley. Protecting free expression online with freenet. *IEEE Internet Computing*, 6(1):40–49, 2002.

120. Freenet. http://freenetproject.org.

121. Ian Clarke. A distributed decentralised information storage and retrieval system, 1999. http://freenetproject.org/papers/ddisrs.pdf.

122. L. Clarke, O. Sandberg, B. Wiley, and T.W. Hong. Freenet: a distributed anonymous information storage and retrieval system. In *Designing Privacy Enhancing Technologies. International Workshop on Design Issues in Anonymity and Unobservability*, 2000.

123. Toby Walsh. Searching in a small world. In *IJCAI*, pages 1172–1177, 1999.

124. Vilhelm Verende. Switching for a small world. Master's thesis, Chalmers University of Technology, 2007. http://www.cse.chalmers.se/~vive/vilhelm_thesis.pdf.

125. Ian Clarke, Oskar Sandberg, Matthew Toseland, and Vilhelm Verendel. Private communication through a network of trusted connections: The dark freenet, 2010. http://freenetproject.org/papers/freenet-0.7.5-paper.pdf.

126. Peter Biddle, Paul Engl, Marcus Peinado, and Bryan Willman. The darknet and the future of content distribution. In *Digital Rights Management: Technological, Economic, Legal and Political Aspects*, volume 2770, pages 344–365, 2003.

127. Salman A. Baset and Henning Schulzrinne. An analysis of the skype peer-to-peer internet telephony protocol. *Proceedings IEEE INFOCOM 2006 25TH IEEE International Conference on Computer Communications*, 6(c):1–11, 2004.

128. David A. Bryan, Bruce Lowekamp, and Cullen Jennings. Sosimple: A serverless, standards-based, p2p sip communication system. In *AAA-IDEA'05*, pages 42–49, 2005.

129. Marti A. Motoyama and George Varghese. Crosstalk: scalably interconnecting instant messaging networks. In *Proceedings of the 2nd ACM workshop on Online social networks*, 2009.

130. Wei Tsang Ooi. Dht and p2p games, 2009. http://www.slideshare.net/weitsang/lecture-7-dht-and-p2p-games-presentation.

131. Adeona, 2009. http://adeona.cs.washington.edu/.

132. Thomas Ristenpart, Gabriel Maganis, Arvind Krishnamurthy, and Tadayoshi Kohno. Privacy-preserving location tracking of lost or stolen devices: Cryptographic techniques and replacing trusted third parties with dhts. In *Proceedings of the 17th conference on Security symposium*, pages 275–290, 2008.

133. Jim Gray. An analysis of the skype peer-to-peer internet telephony protocol. *Queue*, 6(3):63–68, May 2008.

134. David P. Anderson, Jeff Cobb, Eric Korpela, Matt Lebofsky, and Dan Werthimer. Seti@home: An experiment in public-resource computing. *Communications of the ACM*, 45(11):56–61, November 2002.

135. S. Keshav. Efficient and decentralized computation of approximate global state. *ACM SIGCOMM Computer Communication Revie*, 36(1), January 2006.

136. Rafik Makhloufi, Gregory Bonnet, Guillaume Doyen, and Dominique Gaiti. Decentralized aggregation protocols in peer-to-peer networks: A survey. In *Proceedings of the 4th IEEE International Workshop on Modelling Autonomic Communications*, number 1, pages 111–116, 2009.

137. David Kempe, Alin Dobra, and Johannes Gehrke. Gossip-based computation of aggregate information. In *44th Annual IEEE Symposium on Foundations of Computer Science*, 2003.

138. Ji Li, Karen Sollins, and Dah-Yoh Lim. Implementing aggregation and broadcast over distributed hash tables. In *ACM SIGCOMM Computer Communication Review*, volume 35, pages 81–92, January 2005.

139. Marc S. Artigas, Pedro Garcia, and Antonio F. Skarmeta. Deca: A hierarchical framework for decentralized aggregation in dhts. In *17th IFIP/IEEE International Workshop on Distributed Systems: Operations and Management*, October 2006.

140. Matthew Harren, Joseph M. Hellerstein, Ryan Huebsch, Boon Thau Loo, Scott Shenker, and Ion Stoica. Complex queries in dht based p eer-to-peer networks. In *Proceeding IPTPS '01 Revised Papers from the First International Workshop on Peer-to-Peer Systems*, pages 242–259, 2002.

141. Bauer Daniel, Hurley Paul, Pletka Roman, and Waldvogel Marcel. Complex queries in dht based peer-to-peer networks. In *Proceeding 29th Annual IEEE International Conference on Local Computer Networks*, May 2004.

142. Davide Carfi, Massimo Coppola, Domenico Laforenza, and Laura Ricci. Ddt: A distributed data structure for the support of p2p range query. In *CollaborateCom'09*, pages 1–10, 2009.

143. Guanling Lee, Jia-Sin Huang, and Yi-Chun Chen. Supporting filename partial matches in structured peer-to-peer overlay. In *5th International Conference on Grid and Pervasive Computing*, 2010.

144. Tan Yunsong and Wu Yuntao. Efficient range indexing in dht-based peer-to-peer networks. In *International Forum on Information Technology and Applications*, 2009.

145. Guido Urdaneta, Guillaume Pierre, and Maarten van Steen. A survey of dht security techniques. *ACM Computing Surveys*, 43, January 2011.